BIG BANG OUT,
MULTIBANG IN

BIG BANG OUT, MULTIBANG IN

ANDRE TREPANIER

Copyright © 2016 by Andre Trepanier.

Library of Congress Control Number:		2015919768
ISBN:	Hardcover	978-1-5144-3016-3
	Softcover	978-1-5144-3015-6
	eBook	978-1-5144-3014-9

All rights reserved. No part of this book may be reproduced or transmitted in any form or by any means, electronic or mechanical, including photocopying, recording, or by any information storage and retrieval system, without permission in writing from the copyright owner.

Any people depicted in stock imagery provided by Thinkstock are models, and such images are being used for illustrative purposes only. Certain stock imagery © Thinkstock.

Print information available on the last page.

Rev. Date: 11/22/2019

To order additional copies of this book, contact:
Xlibris
1-888-795-4274
www.Xlibris.com
Orders@Xlibris.com

Contents

1. Introduction and Overview..20
   1.1 Introduction 20
   1.2 Problems with Cosmological Models..21
       1.2.1 Big Bang Paradigm..21
       1.2.2 Steady State and Static Universe Tired Light Models...................24
       1.2.3 Occam's Razor and Proposed Model 25

2. Mass Distribution and Density in Cosmos..26
   2.1 Spheres..26
       2.1.1 Free Space in Two Dimensions..26
       2.1.2 Free Space in Three Dimensions (3-D) and Kepler
             Conjecture...29
       2.1.3 Space Reduction between Gaseous Spheres 31
   2.2 Bubbles..32
       2.2.1 Surfaces for Two Bubbles..32
       2.2.2 Junction of Three Bubbles 34
       2.2.3 Tetrahedral Space between Four Bubbles 34
   2.3 Density and Mass Distribution..36
       2.3.1 Density of Stars and their Satellites.................................36
       2.3.2 Galaxy, Cluster, and Bubble Densities 37
       2.3.3 Average Cosmic Density 37
       2.3.4 Average Cosmic Radius and Total Mass..................................39
       2.3.5 Finite and Permanent Cosmos 41
       2.3.6 Friedmann's Equation and Big Bang Model...............................44

3. Electromagnetic Waves, Doppler Effect, and Refraction..........................46
   3.1 Doppler Effect in Daily Life...48
       3.1.1 Doppler Shift Related to Source and Observer 48

        3.1.2   Frequency Shift when Receptor Is Moving away 50
        3.1.3   What Happens when a Balloon Separates the Source and Receptor? 52
   3.2   Cosmic Light Redshifts 53
   3.3   Refraction across a Medium 56
   3.4   Refraction in a Prism 57

4. **Expansion and Redshift, Limited Contraction and Blueshift, Cosmic Microwave Background (CMB)** 59

   4.1   Observation of Data Slices across the Cosmos 59
        4.1.1   Expansion across Bubbles 63
        4.1.2   Maxibangs and Expansion 66
        4.1.3   Light Faster across Bubbles 72
        4.1.4   Light Crossing Time and Expansion 73
        4.1.5   Average Light Trajectory across Cosmic Bubbles 74
        4.1.6   Local Bubble Expansion 76
   4.2   Expansion and Contraction between Bubbles 82
        4.2.1   Dynamic Wall between Two Bubbles 82
        4.2.2   Contraction Space or Blueshift between Three Bubbles 85
        4.2.3   Tetrahedral Space between Four Bubbles 87
        4.2.4   Hexahedral Space between Six Bubbles 89
        4.2.5   Old Bubbles and Dodecahedron 92
   4.3   Expansion and Contraction Intervals 93
   4.4   Distance and Redshift 94
        4.4.1   Redshift and the Moving Sidewalk 95
        4.4.2   Cellular Expansion Model and Redshift 95
        4.4.3   Lookback Time, Distance, and Redshift 97
        4.4.4   Luminosity Distance and Redshift 98
   4.5   Cosmic Microwave Background (CMB) 101
        4.5.1   Black Body Spectra for Three Temperatures 102
        4.5.2   Extragalactic EM Waves Background 104
        4.5.3   Inverse Compton Scattering and the Sunyaev-Zel'dovich Effect 106
        4.5.4   Trace Stacking and Redshift to a Blackbody Spectrum 107

5. **Stars, Galaxies, Groups, Clusters, and Clouds** 111

   5.1   Classification of Stars 112
        5.1.1   Solar Radiation Spectrum 114
        5.1.2   Star Formation 116

	5.1.3	Main Sequence to Neutron Stars and Black Holes............ 118
	5.1.4	Metallicity and Redshift in Astronomy 122
5.2	Families of Galaxies and Clusters 122	
	5.2.1	Spiral Galaxies and Star Globular Clusters 124
	5.2.2	Lenticular and Elliptical Galaxies.................................... 129
	5.2.3	Colliding and Irregular Galaxies...................................... 130
	5.2.4	Active Galactic Nucleus or AGN Galaxies....................... 130
	5.2.5	Intra- and Perigalactic Clouds and Dust 132
	5.2.6	Clusters of Galaxies and Intercluster Medium 135
5.3	Quasar Distribution and Dynamics 138	
	5.3.1	Quasar Distribution... 139
	5.3.2	Quasar Lifetime and Dynamics.. 141
	5.3.3	Quasar Emission and Absorption Spectra........................ 142

6. Dynamics of Galaxies and Clusters of Galaxies.........................149

    6.1 Disk Model and Equations 151
        6.1.1 Disk Model Presentation... 152
        6.1.2 Circle and Disk Equations.. 155
        6.1.3 Cosine-Corrected Gravitational Force............................. 158
        6.1.4 Graphs and Results... 161
        6.1.5 Mass, Velocity, and Distance Ratios................................ 172
    6.2 Application to Four Typical Galaxies.. 173
        6.2.1 Application to a Large Galaxy: NGC 3198...................... 175
        6.2.2 Example from a Small Galaxy: NGC 4062 181
        6.2.3 Application to a Barred Galaxy: NGC 4389 184
        6.2.4 Example from Elliptical Galaxy NGC 3379 188
    6.3 Statistical Results with 46 Galaxies.. 192
    6.4 Orbiting Mass and Spiral Galaxies 198
    6.5 Linear Model for Filamentary Structures in Galaxies,
        Groups, and Clusters 203

7. Four Forces or Facets of Energy in the Cosmos 209

    7.1 Electronuclear Force and Energy Conservation.......................... 209
        7.1.1 Standard Model of Particles 209
        7.1.2 Friction, Entropy, and Energy Conservation 213
    7.2 Gravitation and Electronuclear Rewinding 214
        7.2.1 The Conservative Force... 214
        7.2.2 Gravity and Velocity.. 215
        7.2.3 Cosmic Cyclic Rewinding 217

7.3   Complexity Level and Energy Quality Conservation 225
    7.3.1   Complexity Levels from Gravity to Cosmic Community 225
    7.3.2   Quality Increase and Conservation....................................227
    7.3.3   Winning Cosmic Mix: Gravity, Black Energy,
          Electronuclear, and Biochemical 229

Summary of Key Ideas in Big Bang (BB) and
    Multibang (MB) Models..231

Tables

Table 2.1 Selected bubble and interbubble cosmic parameters 43

Table 3.1 Electromagnetic waves summary 47

Table 3.2 Redshift illustrated from ultraviolet to infrared 54

Table 4.1 Cosmic bubble number and parameters 66

Table 4.2 Fossil walls and extremely low density within bubbles 71

Table 4.3 Parameters used and computed for our local bubble 78

Table 5.1 Stars' spectral classes in terms of various parameters 112

Table 5.2 Cosmic galactic group parameters 132

Table 5.3 Average physical parameters for cosmic clusters and gas in walls ... 136

Table 5.4 Parameters for quasars and maxibangs 141

Table 6.1 List of 30 HSB and 16 LSB galaxies with radius in kpc 193

Table 7.1 Bang time sequence for our bubble 218

Table 7.2 Distribution of growing complexity in cosmos 225

Figures

Figure 2.1 Shortest or minimum trail system between four points..............27
Figure 2.2 Hexagonal or honeycomb pattern..27
Figure 2.3 Circles in a square arrangement 28
Figure 2.4 Hexagonal close packing surface in 2-D 28
Figure 2.5 Face-centered and hexagonal close packing 30
Figure 2.6 Three hexahedrons in the hexagonal close packing...................31
Figure 2.7 Four spheres surrounding a secondary cluster of galaxies............32
Figure 2.8 Minimum surfaces between two bubbles.....................................33
Figure 2.9 Small bubble merging with a larger one......................................33
Figure 2.10 Junction of three bubbles..34
Figure 2.11 Angles and vertices at the junction of four bubbles..................35
Figure 2.12 View above one vertex and facing three of the four bubbles......36
Figure 2.13 Baryon density in M_p/m^3 from nucleosynthesis 38
Figure 2.14 Distribution of density and radius for three models.................40
Figure 2.15 Spheroidal multibang cosmos...42
Figure 3.1 Doppler shift related to fixed or moving emitter and receptor....49
Figure 3.2 Doppler effect when receptor is moving away............................50
Figure 3.3 Expanding sphere between emitter and receptor........................53
Figure 3.4 Comparison between one and two cycles
 for redshift z=0 and z=1 55
Figure 3.5 Refraction across a denser medium 56
Figure 3.6 Refraction of light across a prism 57
Figure 4.1 Two-degree Field Galaxy Redshift Survey....................................60
Figure 4.2 Real and observed cluster 61
Figure 4.3 Small angle slice and projection on the plane..............................62
Figure 4.4 Average spheroidal bubble slice crossing 63
Figure 4.5 Expanding bubble and clusters...64
Figure 4.6 Mature bubbles with quasar in center 67
Figure 4.7 Maxibangs from quasars and expansion 68
Figure 4.8 New walls, galaxies, and clusters..69
Figure 4.9 Time, velocity, and H values for bubble expansion 69
Figure 4.10 Void galaxies in expanding bubble...70

Figure 4.11 Light rays crossing an expanding shell 72
Figure 4.12 Light crossing time and expansion 73
Figure 4.13 Average crossing distance facing the wave 75
Figure 4.14 Light average crossing distance in a sphere 76
Figure 4.15 Local bubble expansion with base at cluster A3526 77
Figure 4.16 Local bubble expansion with base at cluster A3574 79
Figure 4.17 Local bubble expansion with base at cluster A3574 at 0° 80
Figure 4.18 Local bubble expansion with base at Virgo cluster 81
Figure 4.19 Turbulence and galaxy formation between two bubbles 83
Figure 4.20 Wall formation between two expanding bubbles 84
Figure 4.21 Expansion and contraction between three bubbles 85
Figure 4.22 Tetrahedral space between four bubbles 87
Figure 4.23 Cluster in acute tetrahedron 88
Figure 4.24 Active galactic nucleus (AGN) in contracting tetrahedron 88
Figure 4.25 Hexahedron between six spheres 89
Figure 4.26 Concave hexahedron produced by six spheres 90
Figure 4.27 Concave hexahedron and feeders 91
Figure 4.28 Comparison of hexahedron and tetrahedron size 91
Figure 4.29 Dodecahedron or maximum number of bubble walls 92
Figure 4.30 Sketch of light rays crossing many bubbles 93
Figure 4.31 Jim walking in the opposite direction to the moving sidewalk .. 95
Figure 4.32 Self-regulation of growth cells in a 3-D spheroid 96
Figure 4.33 Lookback time or distance in relation to redshift z 97
Figure 4.34 Low-z cosmic time and distance 99
Figure 4.35 Blackbody spectra for three temperatures 103
Figure 4.36 Extragalactic EM Waves Background 104
Figure 4.37 The Sunyaev-Zel'dovich effect across a cluster 106
Figure 4.38 Stack of five signal traces 107
Figure 4.39 Trip of photons redshifted to CMB across the cosmos 108
Figure 4.40 Comparison of 1 mm and 0.001 mm wavelength 110
Figure 5.1 Stars' energy output in terms of spectral class 114
Figure 5.2 Sketch of solar radiation spectrum at sea level and above the atmosphere (Credit: Wikimedia commons) 115
Figure 5.3 Turbulent fluid flow with eddies and whirling cells 116
Figure 5.4 Transparent shell from supernova SNR 0509 (Credit: NASA Hubble Space Telescope) 119
Figure 5.5 Curriculum vitae for different masses and classes of stars 120
Figure 5.6 Galaxy formation between two expanding shells 125
Figure 5.7 Comparison of a hurricane with a galaxy (Credit: NASA/US NOAA and NASA/HST) 126
Figure 5.8 Moments of inertia from four geometrical shapes 127

Figure 5.9 Spiral galaxy evolution from eddy to barred system 128
Figure 5.10 Galactic evolution from low to high density 131
Figure 5.11 Inverted black-and-white galaxy M 83 and surroundings as seen from a range of wavelengths (Credit: NASA/JPL-Caltech/VLA/MPIA) 133
Figure 5.12 Field spiral and surroundings in a wall 134
Figure 5.13 Cluster with a giant elliptical galaxy and a quasar 137
Figure 5.14 Distribution of known quasars from the Vizier catalogue 139
Figure 5.15 Known and projected distribution of quasars 140
Figure 5.16 Quasar Q1323+6530 at z=1.618 and some attributes [86] 142
Figure 5.17 Partial spectrum from quasar Q0014+813 and vertical compaction [87] 144
Figure 5.18 Partial spectrum from quasar Q0003+1553 stretched to the scale at the bottom of figure 5.17 [86] 145
Figure 5.19 Sketch of the average spectrum of a quasar around z=6 with some attributes 146
Figure 5.20 Quasar redshift versus Lyman alpha flux, compared with three inverse laws 147
Figure 6.1 Comparison of gravitational attraction from a sphere and from a disk 150
Figure 6.2 Disk M under study with segment divisions 152
Figure 6.3 Truncated ring computations 153
Figure 6.4 Disk M and variable circle M_n with associated symbols 154
Figure 6.5 Mass computed from volume and density at the edge of disks increasing in size and away from the edge of the largest disk ... 157
Figure 6.6 Cosine angles across the rings and vectors 158
Figure 6.7 Cosine summation for one truncated ring 159
Figure 6.8 Corrected force F_c when unit mass is moved across the disk 160
Figure 6.9 Comparison between cosine-corrected gravitational force F_c across 30 disks. Also, F_c across and away from the largest disk with a radius of 150 units (triangles) and Newton's gravitational force F_n (squares) 161
Figure 6.10 Gravitational force F_c across 30 different disks with radiuses from 0 to 150 units and away up to 200 units in steps of five units 162
Figure 6.11 Ratios: center of mass radius over center-of-attraction radius, R/R_c, at the edge of 30 disks (diamonds), across the disk with radius R=150 (squares), and Newtonian (triangles) 163
Figure 6.12 Top curve is Doppler velocity, V_c, and Keplerian velocity, V_K, is the bottom one 164

Figure 6.13 Comparison of corrected velocity, V_c (squares); and Keplerian velocity, V_K (triangles), for single disk of 150 units 166
Figure 6.14 Example of an orbiting mass A spiraling out on the left and mass B spiraling in on the right 167
Figure 6.15 Inverted colors of low surface brightness galaxy NGC 4921 (Credit: NASA/ESA/HST) 168
Figure 6.16 Corrected mass M (M_c or M_K, squares); apparent mass (triangles) at the edge of 30 disks; and apparent mass (diamonds) across largest disk 169
Figure 6.17 Mass at the edge of 30 separate disks and away as physically calibrated and after model corrections 170
Figure 6.18 Summary of the theoretical model 171
Figure 6.19 Mass ratios, Ma/M, or distance ratios, R/R$_c$, at the top; and velocity ratios, V_c/V_k, at the bottom 172
Figure 6.20 Matrix setup to compute galactic forces, velocities, densities, and masses using 30 disks with R=0 to 150 units in steps of five units 174
Figure 6.21 Black and white image of NGC 3198 as seen by the Spitzer space telescope (Credit to: NASA/JPL Caltech) 175
Figure 6.22 Newton force and corrected force for NGC 3198 176
Figure 6.23 NGC 3198 corrected velocity, V_c (squares); and Keplerian velocity, V_K (triangles) from a stack of 30 disks 177
Figure 6.24 Apparent mass, M_a, on the top and corrected mass, M, at the bottom for NGC 3198 using a 30-disk model. 178
Figure 6.25 Three curves from NGC 3198 corrected or real mass with central spheroid, no spheroid, and no spheroid with 2 kpc intervals. ... 180
Figure 6.26 Mass added and density per ring for NGC 3198 181
Figure 6.27 Newton and corrected force for NGC 4062 182
Figure 6.28 Velocity distributions for NGC 4062 182
Figure 6.29 Apparent and corrected mass for NGC 4062 183
Figure 6.30 Mass added and density per ring for NGC 4062 184
Figure 6.31 Newton and corrected force for NGC 4389 185
Figure 6.32 Velocity distributions for NGC 4389 185
Figure 6.33 Apparent and corrected mass for NGC 4389 186
Figure 6.34 Mass added and density per ring for NGC 4389 187
Figure 6.35 Black and white version of elliptical galaxy 3379 (Credit: NASA/HST) 188
Figure 6.36 Newton and corrected force for NGC 3379 189
Figure 6.37 Velocity distributions for NGC 3379 189
Figure 6.38 Apparent and corrected mass for NGC 3379 190
Figure 6.39 Mass added and density per ring for NGC 3379 191

Figure 6.40 Corrected galactic mass at the edge of selected galaxies 194
Figure 6.41 Ratio between apparent and corrected mass for 46 galaxies 195
Figure 6.42 Average density in M_p/m^3 for the 46 selected galaxies 196
Figure 6.43 Average spiraling in velocity for low- and
high-surface-brightness galaxies 197
Figure 6.44 Average time to spiral in a distance equal to ½ the
circumference at the galactic edge 198
Figure 6.45 Migration of the center of attraction to the center of mass 199
Figure 6.46 Sketch showing the relationship between orbiting mass
and a galaxy 200
Figure 6.47 Distribution of orbiting time and spiraling in time for
NGC 3198 201
Figure 6.48 Linear model used for computations 203
Figure 6.49 Corrected and Newtonian forces for a linear mass
distribution 204
Figure 6.50 Corrected velocity, V_c (squares); and Keplerian velocity,
V_K (triangles). 205
Figure 6.51 Distribution of apparent and corrected mass 206
Figure 6.52 Ratio apparent to corrected or Keplerian mass M_a/M 207
Figure 7.1 Fermions and bosons of the Standard Model 210
Figure 7.2 UCT quarks and projection to Planck's energy 212
Figure 7.3 Sketch of the four laws of thermodynamics 213
Figure 7.4 The fate of Sisyphus rolling forever a boulder up the hill 215
Figure 7.5 Gravitation is the deepest and most overpowering of all forces 216
Figure 7.6 Black hole and maxibang mass, volume, and time distribution ... 219
Figure 7.7 Black hole and maxibang radius with escape
velocity distribution 220
Figure 7.8 Black hole and maxibang density and temperature distribution ... 221
Figure 7.9 Sketch of a heat pump with cold and hot parts 222
Figure 7.10 Sketch of the self-sustained cosmic heat pump 223
Figure 7.11 One complete cycle from maxibang to black hole 224
Figure 7.12 Water molecule formed from three atoms 227
Figure 7.13 A black and white patch of sky from Hubble Extreme Deep
Field (Credit: NASA/ESA; G. Illingworth, D. Magee, and P.
Oesch, University of California, Santa Cruz; R. Bouwens, Leiden
University; and the HUDF09 team) 229

Acknowledgments

This book is dedicated to my Mediterranean wife, Khadija, for her patience and warm presence.

Thanks to my family and all my friends from the class of '63 who encouraged me in writing "Big Bang Out, Multibang In" and especially to Robert Gagnon who made many useful comments on Galactic Dynamics in Part 6.

A special thought for my friends from the foundation set up by the late Roger Lebeuf who wrote a basic philosophical synthesis titled "Cosmic Presence."

Cover picture:

A 400-year-old spherical transparent shell remnant from supernova SNR 0509 taken by the NASA Hubble Space Telescope in the Large Magellanic Cloud, a satellite galaxy to the Milky Way 160,000 light-years away, is shown. A white dwarf or neutron star left over from the progenitor star is visible in the center on figure 5.4, and a shell expansion velocity of 5,000 km/sec has been measured.

Also shown is a 600-year-old polyhedral transparent remnant from supernova SNR 0519 taken by the NASA/ESA Hubble Space Telescope in the Large Magellanic Cloud located 150,000 light-years away. The white dwarf star that exploded at the end of its life was probably similar to our sun.

Preface

A few years ago, as I was still active as a petroleum geophysicist, I followed an introductory course on cosmology given by Hubert Reeves on the education television channel in Québec. The subject was extremely well structured and clearly presented by this expert, but I was astonished to discover the lack of scientific proofs on which the whole castle of modern cosmology is built [21]. Now retired from my field of activity, I have read and reflected for the past few years on this passionate subject.

My goal is to challenge some of the main tenets of present-day cosmology, not as an expert, but simply as anybody questioning some of the shocking conclusions that have been transformed in the course of about one century, from paradigms to pseudoscientific dogmas. The challenge will be conducted in two parts: first, current mainstream ideas will be presented with their acknowledged scientific input and their questionable models and results; second, I will present a defendable alternative model with all due respect to physical laws and show its more universal value and consistency. I hope this original model will stimulate discussions and research in new directions to untangle the Gordian knot choking the prevalent big bang model.

Why should we question such a complex subject where experts keep learning every day, and what could we possibly add that has not already been said? Cosmology is so fundamental that it reaches to our wonderment and deepest feelings about nature and our own cosmic presence. With new technologies and the load of digitized data from ground-based and satellite-based telescopes and accessory instruments, headways have been made with factual data interpretation. But the cosmos is not a private hunting ground and does not belong to a particular club of observers. It is permanently available to anyone who wants to explore it. Like so many scientific ventures, it is based on hypothetical models that are continuously adjusted to fit new observational data. Thus, we are responsible to challenge the current big bang model whose interpretation is generally disconnected from many of the basic observations and propose some improvements where we see fit, all the way from early expansion to final outcome theories.

Our cosmos has some similitude with the black hole where a mass within the Schwarzschild radius is contracted by a gravitational field so intense that almost nothing can escape, not even light. It is also bright since we are observing the universe from within and since it is filled with energy of all kinds that we can tune in. There is no outside, no observer studying our cosmos from a separate vantage point since an outsider would be unable to illuminate it or to capture any of its energy. Thus, our cosmos is an extreme black hole from which nothing can escape, no particle, no photon, no gravity, nothing at all. It is also a perfect blackbody in my multibang model, the ultimate one, a cavity in thermodynamic equilibrium at 2.725 K, which is the temperature of the cosmic microwave background.

So I would like to share with you the following ideas and much more: The unique big bang model is replaced by a multibang model made up of many local and permanently renewed maxibangs. Universal-scale cosmic expansion is challenged by a local-scale expansion and contraction model where expansion is the dominant volumetric expression. Universal cosmological redshift is replaced by local cosmological space expansion redshift, neutral shift, and contraction blueshift which add up to a larger redshift value than the sum of the local blueshifts and neutral shifts, and Hubble's constant is explained as the average rate of expansion computed from a multitude of local bubble redshifts. The cosmic model that will emerge includes time and space as permanently renewed local features of the electronuclear realm within a finite spheroidal cosmos where the conservative force of gravity, freed from the limited speed of light, becomes the great watchmaker rewinding the cosmos.

What are my original new concepts that you may expect to discover in this book? Here are a few of the most fundamental ones:

- It is only with expanding bubbles or cells that the largest structures in the universe may be created and maintained.
- And only with these expanding bubbles can we explain light redshifts coming from faraway galaxies and a different interpretation of Hubble's constant.
- Also, bubbles expansion counteracts the pull of gravity within the cosmos and guarantees stability with homogeneity and isotropy.
- So, Einstein was right-on when he introduced the cosmological constant lambda Λ in his equations of general relativity. It corresponds to bubbles in expansion.

- Dark matter does not exist. It is a concept born from a misinterpretation of star's orbital velocities in spiral galaxies. Doppler velocities must be used with a disk model to calculate proper mass and densities.
- Moreover, corrected vector velocities will unambiguously explain the natural spirals and bars in galaxies, how stars are moving down the spiral arms until reaching a circular orbit and then building up bars on both sides of the central core and bulge.
- You will find that supernovae Ia are true cosmic candles giving directly reliable luminosity-distances in the cosmic ladder, and this with no cosmic expansion and no accelerated expansion, making dark energy a pure ad hoc invention of the big bang model.
- With dark matter and dark energy evacuated from the model, you will find a cosmos filled with 100% baryonic matter generated from extreme black holes or quasars in maxibang events, and with the birth of local time and space.

Therefore, the dynamic cosmos is permanent and in equilibrium between the pull of gravity and the cosmological constant Λ, a constant microwave background temperature of 2.7 degree Kelvin, a constant ratio of different elements.

In the text, I will frequently use the scientific notation in which numbers are traditionally expressed as exponents of base 10; for example, 10^3 is 1,000, where the exponent indicates by how many digits the decimal point in 1.0 is moved to the right, and a fraction like 10^{-3} is 0.001, where the exponent indicates how many times or digits the decimal point in 1.0 is moved to the left. Note that this is different from most handheld calculators' notation, where numbers are expressed as exponents of 1, like 1^3 is 1,000, and also contrary to computer spreadsheets' notation where the more explicit value 1E+3 representing 1,000 will be used in data tables.

Although not specifically shown, force, velocity, and acceleration are vectors with magnitude and direction, whereas time, mass, and distance units are scalars expressed with magnitude only. References to authors or articles listed in the bibliography are shown with bracketed numbers [00] in the text. Even though some sketches have been inspired and modified from published data, all figures and tables have been prepared by the author using the free Paint.NET software, with the exception of 12 figures: 4.1, 5.2, 5.4, 5.7, 5.11, 5.16 to 5.18, 6.15, 6.21, 6.35, and 7.13 for which proper credit is mentioned in the text. Anyone interested to share comments on this book or to receive the Excel sheets used to compute the theoretical model and its application to evaluate corrected galactic velocities and masses is welcome to ask a free e-copy from the author at: atkb22@gmail.com

1. Introduction and Overview

Let us start this cosmic exploration with a brief historical review and a summary of the main theories and associated problems with present-day cosmological models, including the prevailing big bang paradigm and the less-publicized steady state or static universe models [14]. We will apply Occam's razor principle throughout the text to facilitate the understanding of our own model and consequently present it with the least assumptions.

1.1 Introduction

Vesto Slipher may be considered as the father of galactic Doppler shift measurements, which he started in 1912 on the so-called nebulae since separate galaxies were not yet identified. He later discovered that most of these nebulae were redshifted and thus receding from earth. At that time, galaxies were known from the 100-year-old Messier catalog as nebulae, and since the cosmic ladder of distances was not yet invented, it was generally believed that these nebulae were within or in proximity to the Milky Way.

In 1922, Alexander Friedmann derived his own equations from Albert Einstein's equation of general relativity from which he showed that the universe could be expanding although a static universe was the privileged model in those days and it would be difficult to convince the scientific community otherwise [62].

Only two years later, Edwin Hubble started measurements on Cepheid variable stars and determined that galaxies were located at great distances outside our Milky Way galaxy. In 1927, Georges Lemaître presented the idea that the recession of the "nebulae" found from redshift is due to the expansion of the universe. Then in 1929, Hubble discovered the well-known Hubble's law based on redshift that correlates distance and recession velocity although the original version was a very poor correlation with a large error in distances. Since the universe may be expanding, Lemaître suggested in 1931 to go back in time to discover the birth of the cosmos from a small point, a "primeval atom," marking the beginning of space and time.

On the other hand, Friedmann proposed an oscillatory universe model with expansion followed by contraction. Fritz Zwicky came up with the tired light model in which light traveling long distances loses gradually some of its energy to the medium and thus appears redshifted. The latter will be discussed as a static universe non-expansion alternative to the steady state model.

After World War II, three main trends emerged to explain universal expansion or static model associated with redshift. The first was Zwicky's renewed tired light model by followers defending a static universe. The second was put forward by Fred Hoyle and consisted in a steady state model where minute amount of matter are continuously created to compensate for the expansion and thus keep a constant cosmic density. The third was the big bang theory proposed by Lemaître and developed by George Gamow, who introduced the concept of nucleosynthesis, and also by Ralph Alpher and Robert Herman, who predicted the microwave background radiation, whose discovery in 1964 strongly favored the big bang as the better of the two theories to explain the origin and evolution of the cosmos. From that time until now, the vast majority of investments, research, and publications have been directed at the development and defense of the big bang model since no viable alternative has been presented. It is my objective to present such an alternative.

1.2 Problems with Cosmological Models

Big bang, steady state, and static universe models bring some valuable input to the cosmological picture, but they are all riddled with flaws, including some unjustified infinite values [20]. The big bang model has too many mysterious and improbable solutions that have been added every time a problem comes up with observation. The second model, steady state, would require the incredible continuous creation, out of nothing, of new matter and just in the right proportions to fit the actual count as the universe expands. The tired light static universe alternative would require cosmic densities thousands of times above actual measurements to produce the observed redshift [16].

1.2.1 Big Bang Paradigm

When we extrapolate the big bang to its genesis at a theoretical zero second, we are faced with the sudden appearance of a relativistic singularity which involves infinite density and temperature at a finite time around 14 billion years ago [52] or the cosmic creation out of nothing. The defenders of the model will assert that we cannot go beyond the Planck epoch at 10^{-43} second when a blank occurs in physical laws that are no longer applicable, so this qualifies as a first unexplained mystery. The period between the Planck epoch and 10^{-3} second is speculative and unresolved because energies exceed our human experimental capacities limited to about 10^{12} K, but this period also includes the mysterious inflationary epoch around 10^{-35} second, where the universe should have expanded exponentially at a speed much higher than that of light in order to save the isotropic and homogeneous uniformity as stated by the cosmological principle.

The model involves rapid growth and cooling with the passage from ultradense energy to quarks and neutrinos that will mute into matter-antimatter pairs with a positive segregation in favor of matter that will thereafter dominate this baryogenesis and be represented by protons, neutrons, and electrons. Within a few minutes and at a temperature of about one billion K, hydrogen nuclei, lithium, deuterium, and a large fraction of helium were formed and had survived in a process called nucleosynthesis. According to the cosmological model, lithium nucleons cannot survive in most stars since they are transmuted to helium above a temperature of 2.4 million K, but colder brown and orange dwarf stars can keep their lithium fraction. Similarly, deuterium is destroyed very quickly in most stars by fusion into helium at a temperature as low as one million K but could be preserved from burning in colder bodies. Or could the two latter elements be generated in stars or hot gas clouds at about one billion K and be preserved by a rapid cooling process? The presence of about 25% helium-4 by mass or 8% by the number of atoms in the universe is also accounted for by the nucleosynthesis model since stars cannot have produced this large amount of helium-4 from hydrogen fusion in the time frame since the big bang, but it could if a longer time period were available.

Around 379,000 years after the big bang, hydrogen and helium atoms are formed by the capture of electrons, and light is free to escape and supposed to form the cosmic microwave background (CMB) radiation, but since the velocity of light is far superior to expansion velocity, the question is, on what reflector will this escaping light bounce back to become the microwave background about 13.7 billion years later? How will it reach us? After such a big bang, we would expect to observe the equivalent of a large supernova remnant: a cosmos forming a huge spherical shell structure on which we would be living and with the original light gone far away from us forever. This is a questionable concept indeed.

The big bang proponents, assuming that space is like a stretchable material, invented a special space-expansion metric based on relativity called FLRW (Friedmann-Lemaître-Robertson-Walker) to explain space-expansion redshifts. The big bang model seems to be right on the general fact that cosmological redshifts from space expansion may be interpreted as a Doppler shift corresponding to the equivalent of a recessional velocity from which distances may be evaluated and a cosmic distance ladder may be built, although Hubble's law based on a linear relationship between velocities and distances is still debated for the nearby universe.

We will see later that cosmic redshift is the sum of a large number of specific cosmological space redshifts, neutral shifts, and blueshifts from expanding, stable, or contracting cells or bubbles across the cosmos and should be valid as a measure of distances on a large scale if we correct for blueshifts, neutral shifts, and average redshifts, but it does not imply a unique and central universal expansion. The Copernican principle, stating that we are not in a privileged location near the center of the universe, is an acceptable concept since no peculiarities were found about our galactic environment compared with other surroundings, so we could be located anywhere in the cosmos. But on the contrary, the cosmological principle of homogeneity and isotropy, which is also established from observation, may be invalidated in a metric expansion model from a central big bang since its unique expanding shell is not an isotropic or homogenous spheroid.

An apparent mass excess was measured in the 1970s from a large number of rotation velocities computed from redshifts at the edge of galaxies or from clusters of galaxies [35]. This led to the conviction that about 90% of cosmic mass is in the form of nonbaryonic dark matter, which was never observed to this day. Moreover, an interpretation of accelerated expansion from 1A supernova redshifts resulted in the hypothesis that about 70% of the total cosmic mass is in the form of exotic dark energy, for which there is absolutely no tangible proof but for which the cosmological constant was resurrected from Einstein's general theory of relativity. The problem is that this cosmological constant is 120 orders of magnitude smaller than the value expected from quantum gravity theory. We will demonstrate that these exotic dark energy and nonbaryonic dark matter can be excluded from a consistent cosmic model and that the finite universe is closed, thus excluding all flat or open hyperbolic models.

There are also major problems in the big bang paradigm to explain large-scale structure formation and evolution since clear pictures from afar are not showing the predicted evolution from a young cosmos born 13.8 billion years ago. Observations cannot prove that the cosmos was hotter and denser in the past, or that it contained fewer metals, or that the CMB originates about 379,000 years after the bang. All we can see and corroborate is a uniform cosmos made up, at large scale, of expanding bubbles at various stages of dynamic expansion and each one with its own evolving age, temperature, density, metallicity and history.

1.2.2 Steady State and Static Universe Tired Light Models

In the steady state model developed by Hoyle and associates in 1948, the universe is infinite in time: no beginning and no end. Instead of a one-shot big bang creation at the beginning, matter is steadily created to compensate for the decrease in density due to expansion, and thus, the cosmological principle of homogeneity and isotropy is perfectly respected. The amount of new matter needed is about one hydrogen atom per cubic meter per billion years, and also needed is a mass of exotic dark matter five times that of baryonic matter. To this end, Hoyle proposed a "creation field" (called C-field) with negative pressure to act against gravity and produce the measured expansion. The model was later modified by replacing continuous creation with discrete minipockets of creation.

The first major problem is that, with infinite time, we expect infinite accumulation of by-products from processes like nuclear fusion, and the second major problem stems from the infinite material creation in infinite space expansion, which does not make sense in a cosmos filled with finite units of material. Since no mechanism for primordial nucleosynthesis of lithium, deuterium, and helium is known in stars, this model would imply the unlikely continuous creation in the right proportion of those primordial elements. Also, in this model, the microwave background radiation is explained as light from old stars that is scattered in dust. The problem is that grains of dust vary extremely in size, temperature, distribution, and scattering properties, so we should not find the smooth microwave background that is observed as a fact.

Zwicky's tired light model in a static universe with no expansion is the subject of renewed interest and discussions in recent years [27]. Among others, Paul Marmet wrote many papers where he explained that semiclassical electrodynamics and quantum electrodynamics theories consider all interactions between photons and atoms as inelastic, which imply photons losing some energy. The basic idea is that photons traveling across the cosmos are not subject to a cosmological Doppler redshift but are gradually losing an infinitesimal part of their energy, about 10^{-13}, every time they interact with an atom or molecule and that the energy lost is proportional to measured redshift. But the problem is that an average density of about 10,000 hydrogen atoms/m^3 is necessary to produce this effect. This density represents about fifty thousand times the density of the cosmos as measured from star and gas density and from primordial nucleosynthesis theory. It is so large that the universe would rapidly collapse in parts and in totality. Also, the tired light model does not have a cosmological constant or the equivalent to counteract the pull of gravity and keep the universe static. Moreover, we can show that energy is conserved while waves are stretched by redshifting.

1.2.3 Occam's Razor and Proposed Model

William of Occam was an English Franciscan friar who lived in the 14th century and to whom is attributed the principle of using the razor to shave off extraneous assumptions and to use the smallest number of hypotheses to solve a problem. It is not a proven law or theory but an approach to science and knowledge where simplifying usually brings a better result, but not always. It is a general guiding rule in the development of models in physical sciences that will be applied heuristically to the proposed model in this writing.

The basic geometry of the cosmos will be introduced in the shape of spheres and bubbles from which the cosmic mass distribution will be established for our spheroidal universe. Cosmological redshift will be presented as associated with every local bubble expansion and not as a total universal expansion. We will follow some of the light travels with refraction and interaction across cosmic clouds, filaments and other medium, all the way to various backgrounds, including the cosmic microwave background.

The origin of every bubble expansion will be rooted in active galactic nucleus and quasar contraction followed by a maxibang as the source of local baryogenesis and nucleosynthesis, and of a new, expanding shell of material that will act as an equivalent to the cosmological constant lamda opposing the effect of gravity, as predicted by Einstein, and locally creating the next generation of stars, galaxies, and clusters. The apparent mass excess in galaxies and clusters of galaxies will be explained by real and apparent velocities related to the new expansion dynamics and to non-Newtonian centers of gravitational attraction, but respecting Newton's universal law of gravitational attraction. So exotic dark matter and dark energy will no longer be needed since no mass excess and no accelerated expansion is present. Finally, the four forces or bonds of energy in nature will be presented as the gravitational, the black, the electronuclear and the biochemical.

2. Mass Distribution and Density in Cosmos

After a brief look at efficient minimum distances in a two-dimensional surface pattern and a volumetric review of solid spheres in terms of a geometrical arrangement in two or three dimensions, we will see the spatial distribution of expanding bubbles, walls, and junctions between bubbles. The mass and density of stars, clouds, filamentary structures, galaxies, and clusters will be presented in order to establish an average cosmic density, an average radius, and the total mass of a finite cosmic model.

2.1 Spheres

A sphere represents a basic geometrical shape in nature for which we find the smallest minimum surface for a given volume. Spheres are found at all scales and levels of complexity, from the atomic and molecular scale to complex organism like cells, and they are most important in astronomy at small and large cosmic scales.

2.1.1 Free Space in Two Dimensions

Before discussing spheres, let us have a look at some efficient two-dimensional surfaces. What is the shortest trail or minimum road system between four points forming a square whose side has a length of one unit in two-dimensions (2-D)? Looking at figure 2.1, we see that distance AB=BD=DC=CA=1, so the sum of the four sides is equal to four. It is proven that the network, shown in the figure with angles of 120° between every two segments around points E and F or triple-point junctions, yields the minimum value of $1+\sqrt{3}$ or 2.73 units, when we add all the segments. This is much less than for the square where a sum of four was found. In fact, it is about 4% less than the next minimum value of $2\sqrt{2}$ or 2.83 found when we directly connect A with D and B with C as a crossroad system.

Figure 2.1 Shortest or minimum trail system between four points.

This pattern of minimum length on a surface, also shown in figure 2.2 and associated with angles of 120° around triple-point junctions, is frequently found in nature and represents a way to use a minimum amount of energy. For example, bees make their honeycomb in this fashion with a minimum of wax. But how do they measure the 120° angles as they gradually build the comb from top to bottom is still an unanswered question. Is it likely some kind of innate behavior, like the dance they perform, to indicate distances and directions to new flower fields to fellow workers? Also, triple points or hexagonal patterns are found in frost and snow crystals since they are made up of water molecules with angles close to 120° (actually 104.45°).

Figure 2.2 Hexagonal or honeycomb pattern.

If we pile up circles of equal size, for example, disks, on top of one another in a square or rectangular arrangement, we end up with something looking like figure 2.3 shown in two dimensions (2-D). Using a radius of one unit, we find a surface of π, or 3.1416 for each circle enclosed in a square which has a surface of four, so the circular surface is only 79% of the total and the empty surface is about 1/5 of total at 21%. But this is not an efficient way to occupy the space available with identical circles and minimize the empty area.

Figure 2.3 Circles in a square arrangement

For a two-dimensional surface array, figure 2.4 is showing that the hexagonal packing, where a circle is tightly surrounded by six equal circles, will leave much less space between circles. This pattern can be repeated on and on to cover any surface. J. L. Lagrange proved in 1773 that this lattice arrangement has the highest density corresponding to $\pi/\sqrt{12}$ which is about 91% of the surface. This leaves only 9% of the surface for all empty areas located between any three circles and corresponding to those acute triangles with three angles less than 60° each for a sum of angles less than 180°.

Figure 2.4 Hexagonal close packing surface in 2-D

When we compare with figure 2.2, we realize that the same horizontal and diagonal efficient lattice arrangement is present: one showing hexagons and the other showing circles.

2.1.2 Free Space in Three Dimensions (3-D) and Kepler Conjecture

When we look at three-dimensional (3-D) space instead of 2-D, we discover different space fillings. Imagine figure 2.3 as a 3-D drawing of spheres in a boxy or cubic arrangement. Then we would find that each sphere with a radius of one unit represents a volume of $4\pi/3$ or 4.19, which amounts to 52% of 8, the total cubic volume, and that the empty or void space around the sphere represents 48%, which is almost half the total volume. But we will see that the Kepler conjecture offers a more efficient packing of spheres with minimum free space in between. It corresponds to the way spherical fruits like oranges are usually and naturally stacked up in a grocery store, starting at the bottom with a compact hexagonal lattice as in figure 2.4.

The Kepler conjecture has only been proven in 1998 by Thomas Hales. It is a 400-year-old geometric conjecture stated by astronomer Johannes Kepler who wrote in 1611 about the packing of spheres in 3-D space [64]. He asserted that we cannot fill space with a greater density than these two equivalent arrangements of equally sized spheres: the first being a face-centered cubic close packing and the second being called the hexagonal close packing as shown in figure 2.5. Note that three spheres have been separated from both the top and the bottom of the two arrangements to give a better view and that the seven central spheres are placed in the same hexagonal order as the 2-D example of figure 2.4. The difference between the face-centered and the hexagonal close packing is the rotation of the top three spheres with an angle of 60°, which gives a third layer to the face-centered arrangement. We will see later that this rotation has an important consequence in the arrangement of free space for galaxies in a cosmological model. The face-centered cubic arrangement is named from its 3-D lattices forming cubes where each face is made up of five spheres, like letter X with four spheres in the corners and one at the center. The latter has three different layers, or planes, that repeat themselves like ABCBA. But the hexagonal arrangement has only two different layers repeating themselves on top of one another like ABAB.

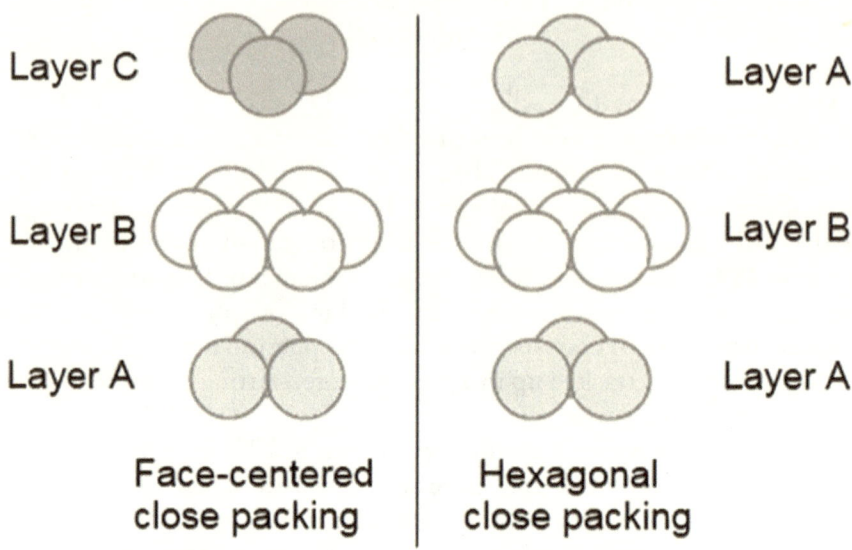

Figure 2.5 Face-centered and hexagonal close packing

The close packing sphere density of both arrangements is $\pi/\sqrt{18}$, which is equal to 0.74 or 74% of the total volume, so that maximum empty space left to be filled with filamentary structures and clusters of galaxies is 26% if, of course, the cosmos could correspond to such a perfect and rigid model. We observe the following facts for the spheres involved in these arrangements:

- Each sphere is in contact with 12 other spheres from which a dodecahedron may be formed.
- A triangular space with three acute angles less than 60° is present at the junction of any three spheres as shown in figure 2.4.
- An opened tetrahedral volume is present at the junction of any four spheres, so each sphere may be attributed ¼ of that space.
- Each sphere shares eight tetrahedrons with other spheres.
- So eight tetrahedrons shared by four spheres each gives 8/4=2 tetrahedrons per sphere, which could correspond to two secondary clusters of galaxies per sphere on a cosmic scale.
- The distance between two tetrahedral volumes is equal to the length of the filament or duct formed by the triangular space between any three spheres and measures approximately 15% more than the spherical radius or less if compaction is present.

- There are also six hexahedral openings associated with each sphere. Three hexahedrons are shown in figure 2.6.
- Since each hexahedron is surrounded by six spheres, this yields one hexahedron per sphere; these are the primary large openings where gas and galaxies will gather and form large clusters of about one thousand galaxies plus associated gas that will gradually be compressed to extreme black holes and produce the future bangs.

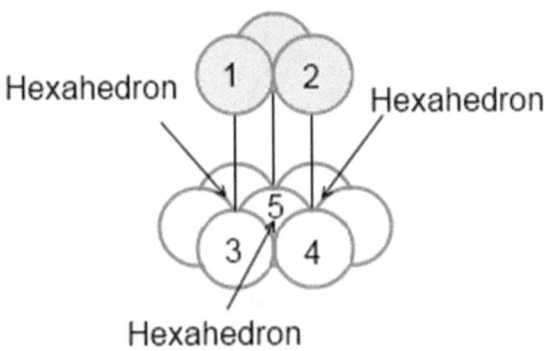

Figure 2.6 Three hexahedrons in the hexagonal close packing

In figure 2.6, we indicate with three vertical lines the three tetrahedral cavities that will be covered by the bottoms of the three upper spheres, and we also show where the three top hexahedrons are located in a hexagonal close packing arrangement. The three bottom hexahedrons are not shown but located directly below the three top ones, for a total of six around sphere number 5. Only five of the six spheres forming the front hexahedral space are indicated since the sixth one, up front, would hide the others. An important relationship to notice is that in the face-centered cubic close packing, the top hexahedrons are rotated with an angle of 60° so that they are not located directly on top of the bottom ones, but they are staggered and form a different plane.

2.1.3   Space Reduction between Gaseous Spheres

Four spheres in a close packing contact are shown surrounding an opened tetrahedral volume in figure 2.7 where a cluster could be nested with filaments coming out from the four triangular openings between any three spheres.

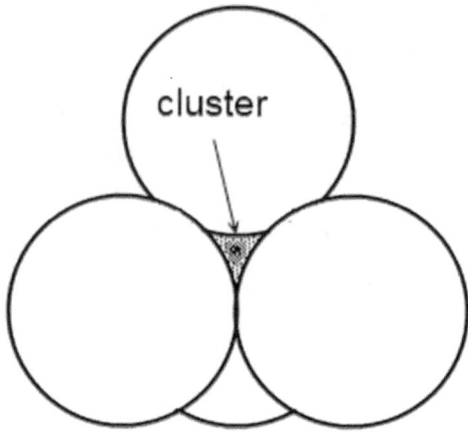

Figure 2.7 Four spheres surrounding a secondary cluster of galaxies

But we are not dealing with rigid spheres in intergalactic and intercluster space. On the contrary, we find decompressing and cooling gaseous walls with galaxy-bearing filamentary structures surrounding large, extremely low-density spherical and ellipsoidal structures with flexible shapes. So the cosmos looks more like a packing arrangement where grapefruits, oranges, tangerines, and kumquats are piled up in a random fashion and partly squashed.

2.2 Bubbles

We will now go from rigid spheres to more flexible spherical forms and look at the shape and special characteristics of some bubbles, especially when we combine two, three, or four bubbles to form common walls. Soap bubbles represent physical examples of complex mathematical problems related to minimal surfaces [75].

2.2.1 Surfaces for Two Bubbles

As shown in figure 2.8, when two bubbles of equal size join together, they form a common wall with a flat surface because water mixed with soap, for example, decreases the water surface tension and thus naturally helps to keep the liquid film from breaking and minimizes the areas for the enclosed volumes. So this simple double-bubble model contains three different surfaces made up of two partial spherical minimum surfaces for the enclosed volumes and a flat disk forming the common wall.

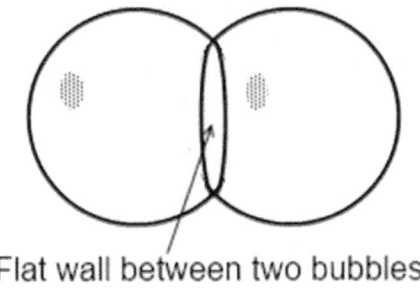

Flat wall between two bubbles

Figure 2.8 Minimum surfaces between two bubbles

It is only in 1995 that Joel Hass and Roger Schlafly announced a proof of this conjecture, stating that three spherical surfaces are formed when two bubbles of equal volume coalesce, and the central surface being flat will form angles of 120° with the curved surfaces and, in this special case, a radius equal to $(R\sqrt{3})/2$ or about 87% of bubble radius R.

On the other hand, the proof that two merged soap bubbles of unequal size, as in figure 2.9, yield the best way of enclosing two different volumes of air with the minimum surface area is still open for proof.

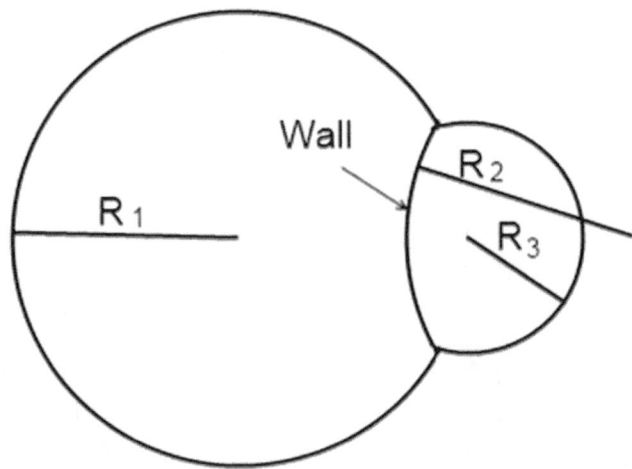

Figure 2.9 Small bubble merging with a larger one

The greater surface tension, and thus pressure, in the smallest bubble produces the incurving inside the larger bubble. Nevertheless, if the double-bubble conjecture applies to this asymmetric case, as shown on empirical ground, then the radiuses of the three spherical surfaces must satisfy the following relationship in which the radius of the two bubbles are R_1 and R_3, and the radius of the common surface is R_2, yielding this equation:

$$1/R_1 = 1/R_3 + 1/R_2 \qquad (2.1).$$

For example, if $R_1=1$ and $R_3=½$, then $R_2=-1$.

2.2.2 Junction of Three Bubbles

Figure 2.10 shows a 2-D sketch representing three coalescing equal-volume and equal-surface bubbles that are each separated from their spherical neighbors by two flat surfaces.

Figure 2.10 Junction of three bubbles

These flattened surfaces form three angles of 120° around the central line joining all three surfaces.

2.2.3 Tetrahedral Space between Four Bubbles

The first scientific experiments with soap bubbles were initiated by Joseph Plateau in the 19th century. He was the first to demonstrate physical solutions to the minimization problems using liquid soap with a variety of geometrical shapes and polygons. This experimenter stated three geometrical conditions always met by soap films bounding any frame:

- The three smooth surfaces of a film intersect along a line, as in figure 2.10 and 2.11.
- The angle between any two surfaces at any point along the line of intersection of three surfaces is 120°, as in figure 2.10 and 2.11.
- Four of the lines, each formed by the intersection of three surfaces, meet at a single point, and the angle between any pair of adjacent lines is 109.28°, as shown in figure 2.11.

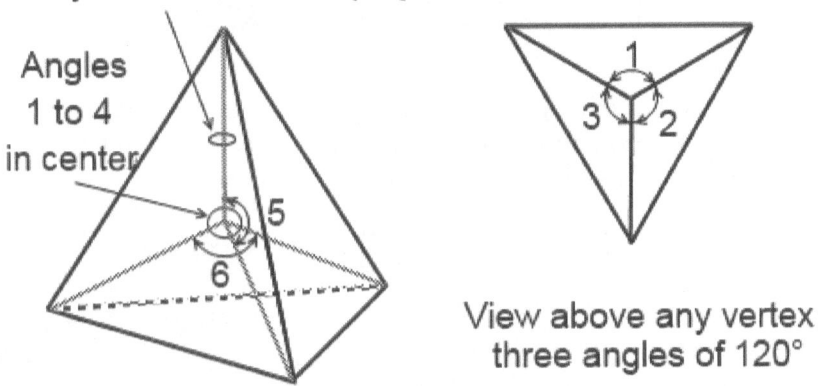

Figure 2.11 Angles and vertices at the junction of four bubbles

We can view this meeting of four bubbles at the center of the tetrahedron if we visualize the four spherical portions of the bubbles outside each face of the tetrahedron as seen in figure 2.12 with the four vertices indicated. The dotted lines in the background show the fourth bubble and the base of the tetrahedron joining the vertices.

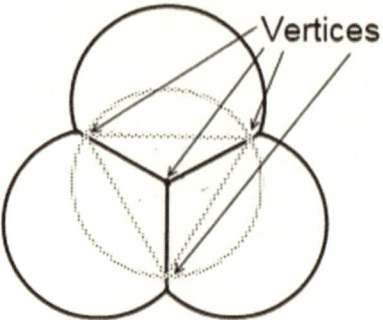

Figure 2.12 View above one vertex and facing three of the four bubbles

This is the same view above one vertex as on the right of figure 2.11, where the spherical part of the bubbles was absent. Note that a wall is present between any two bubbles for a total of six separate walls.

2.3   Density and Mass Distribution

The permanent and finite cosmos is filled with energy and knots of energy observable from a variety of qualitative and quantitative characteristics. Now let us make an assessment of densities and associated masses at cosmological scale, including stars and their satellites, filamentary gas structures and galaxies, clusters and their nested hot gas, and low-density bubbles or voids with their sparse galactic and gaseous inhabitants. Note that although *void* is a good term to represent the extremely low density of a large cosmic volume, the term *bubble* will be used most often for its more dynamic and punchy significance [43].

2.3.1   Density of Stars and their Satellites

Here on earth, we live in a high-density medium where the three phases identified as gas, liquid, and solid, added to a touch of plasma, exchange their energy in a relative equilibrium. Air, for example, with the lowest density of the three phases, is a mixture of gas with a density of 1.2 kg/m^3 at sea level for a temperature of 21°C. The average earth density is 5,515 kg/m^3, similar to other rocky planets and satellites in the solar system. Our sun, like other stars, is mainly composed of plasma, where atoms are stripped of their electrons, more so toward the radiating center, where densities and temperatures reach atomic fusion levels. Its average density of 1,408 kg/m^3 is only 41% more than water density of 1,000 kg/m^3 but varies enormously between the corona at 1.0×10^{-12} kg/m^3 and fusion center at about 1.62×10^5 kg/m^3. Hydrogen nuclear fusion stars have a range of masses between about 8.5% of solar mass for the subdwarfs and more than 100 times the solar mass for the hypergiants, with proportional density variations. If we exclude hypothetical massive singularities and black holes, the champions of density are the gravitationally collapsed massive stars called neutron stars, with a compacted mass of 1.35 to 2.0 times the solar mass in a sphere with a radius of about 12 km and a staggering density up to 6.0×10^{17} kg/m^3, which is twice the atomic nucleus density.

2.3.2 Galaxy, Cluster, and Bubble Densities

If we compute the average density for the equivalent of 160 billion solar masses in our Milky Way galaxy, using a radius of 65,000 light-years, expressed hereafter with the symbol l-y, and a thickness of 1,000 l-y, we find a relatively low value of about 2.8×10^{-20} kg/m^3, which may also be expressed more conveniently as 17 million M_p/m^3, where M_p means "proton mass" or about the mass of one hydrogen atom. Then the next step up is represented by groups of about 40 galaxies where a radius of 5 million l-y in a spherical volume yields a much lower density value of 3.34×10^{-27} kg/m^3, which is equivalent to 2.0 M_p/m^3. Clusters of galaxies are the largest-known massive structures in the universe. One cluster has an average radius of about 13 million l-y and contains an average of 500 galaxies, which gives, in a spherical volume, an average density of 2.0×10^{-26} kg/m^3 or 12 M_p/m^3. On the other hand, spheroidal or ellipsoidal expanding bubbles are the largest low-density structures, much larger than clusters of galaxies: with an average radius of 91 million l-y, their average density of 3.9×10^{-29} kg/m^3 or 0.023 M_p/m^3 is about 10 times less than the universal average of 0.21 M_p/m^3. The presence of 5% of the galaxies in residual walls and filamentary structures within the bubbles has been included in the balance sheet.

2.3.3 Average Cosmic Density

About 13.8 billion years ago, our vast and entire local northern bubble (more visible and more studied from the more populated northern hemisphere on earth), estimated to have a present radius of 100 million l-y, including all related galaxies, neutrinos, and electromagnetic energies, was compressed into an extremely small volume and bounced back on a gravity wall. Like for the Phoenix reborn from its own ashes, this maxibang event represents the birth of our northern bubble out of the ashes of previous galaxies and other captured and compacted masses and energies. This birth, after a gravitational compaction and bounce back, marks the beginning of our local space and time. Of course, there is also a southern bubble with a similar age and history that is located on the other side of the common wall in which our galaxy and close neighbors are embedded.

A chronology of the first instants into each bubble expansion, and assuming similar cosmic results to the one described by big bang model, is expected in the multibang model and transposed to every maxibang in the cosmos. This includes that period of expansion between about three and 20 minutes when nucleosynthesis occurred at a temperature of about one billion Kelvin (K) and a density, like air on earth, of about 1.2 kg/m³ or 7.2×10^{26} M_p/m³. As expansion and cooling took place, protons survived the anti-protons' mass destruction and formed hydrogen nuclei. Figure 2.13 presents a summary of some published data on the average abundance of helium-4, helium-3, deuterium, and lithium-7. These ratios are calculated, as shown on the graphs, to be 12.5 hydrogen atoms for each helium-4, 30,000 hydrogen for each deuterium atom, 90,000 hydrogen for each helium-3, and 5.0 billion hydrogen atoms for each lithium-7. The result indicates a rough agreement between all four values in which deuterium is a good match and helium-4 is an acceptable match, but there are important uncertainties for helium-3 and lithium-7. Since stars cannot produce this large amount of helium-4 in the time frame of our bubbles, it sets a constraint on the model [51].

Figure 2.13 Baryon density in M_p/m³ from nucleosynthesis

All of these abundances are related to a single parameter, which is the ratio of photons to baryons calculated from cosmic microwave background fluctuations and which is not expected to vary with time. So all four nucleosynthesis elements could have a common photon-to-baryon ratio shown by the dotted vertical strip centered on about 1.65×10^9 photons per baryon and ranging in baryon density from 0.18 to 0.24 M_p/m^3 from which we will use the average of 0.21 M_p/m^3 for the entire cosmos.

This density is separately confirmed since it represents also the sum of two values measured by physically adding the baryonic material in the universe: a surprisingly small 0.03 M_p/m^3 from the stars and their satellites and 0.18 M_p/m^3 from gas clouds and hot gas trapped in clusters. The latter two values add up to 0.21 M_p/m^3 and also confirm the ratio of six times more gas than star baryonic material to be applied in large-scale cosmic computations. Evaluations of neutrino and electromagnetic mass density input to the total cosmic count generally indicate negligible values. For example, the cosmic background radiation corresponds to a tiny mass density of 0.0003 M_p/m^3.

2.3.4 Average Cosmic Radius and Total Mass

From Kepler's and Newton's equations, we can find the velocity of a satellite with a negligible mass in a circular orbit around a more massive and spherically symmetric or point-like central body using the well-known equation

$$V = \sqrt{GM/R} \qquad (2.2),$$

where V is the circular velocity, G is the gravitational or Cavendish constant, M is the mass of the massive body and R is the circular radius of the orbiting satellite. This equation yields density and radius for the middle curve in figure 2.14, where the horizontal scale runs from a radius of 10 to 200 billion l-y and the vertical scale covers densities from 0.01 to 100 M_p/m^3. But if we multiply by two the term under the square root in this circular orbit equation, we obtain the escape velocity or Schwarzschild equation when V is replaced by the speed of light c

$$V = \sqrt{2GM/R} \qquad (2.3)$$

from which the density and radius is extracted and shown as the bottom curve in figure 2.14.

However, in a closed and finite permanent spherical universe, we do not want any particle or any kind of energy to escape, not even photons, so the Schwarzschild equation is unacceptable and rejected since some of the mass energy would definitely escape in this case.

On the other hand, equation (2.2) is telling us that some energy or matter may also escape even though it would be limited to a circular orbit around the central mass. The problem is, when we use the fastest velocity, which is the velocity of light, we can expect the accumulation of photons and particles overcrowding the circular orbits around the central mass, so this option must also be rejected since we don't observe a leaking and an evaporating cosmos. Thus, in order to keep a lid on the cosmos and avoid any escape or collapse of mass energy from multipole irregularities, we need to divide the right-hand side of equation (2.2) by a value larger than one. To have a safe and stable model, we may use a value of two to obtain the following modified equation:

$$V=\sqrt{GM/2R} \qquad (2.4).$$

This is the equation that is applied in figure 2.14 for the top curve with the triangular symbols. The selected density of 0.21 M_p/m^3 is reached as indicated by the arrow at a radius of 143 billion l-y. This cosmic radius is about 10 times more than big bang estimates, although the density of 0.21 M_p/m^3 for ordinary baryonic matter is the average predicted by the big bang nucleosynthesis model.

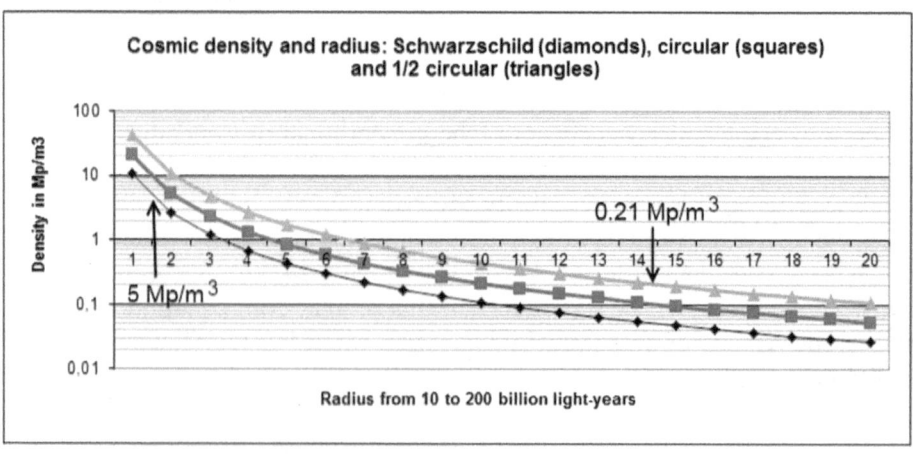

Figure 2.14 Distribution of density and radius for three models

Also, note that for earlier big bang models, before the reintroduction of the cosmological constant, a value of 5 M_p/m^3 is shown with an arrow pointed at the Schwarzschild curve, which indicates the approximate value of the critical density that makes the cosmos stands on the edge between expansion and contraction. But more recent big bang models of the Friedmann universe have added hypothetical dark matter, whereby the actual density is equal to the critical one, and also hypothetical dark energy to explain that the universe is not expected to contract but seems to accelerate its expansion so that the latter should last forever and gradually end up in a dissociated and completely destroyed, cold, dead cosmos when all the electronuclear energy is dispersed and consumed [56].

2.3.5 Finite and Permanent Cosmos

What is the topology of the cosmos? With a known radius and a permanent spheroidal model, what does the cosmos look like? Figure 2.15 is a 2-D representation of the cosmic multipole spheroid filled with bubbles and clusters and where particles and photons come grazing near the edge without ever escaping and without reaching a circular orbit either. Particles will fall back more quickly than light will and in proportion with their respective speed near the edge, as more massive particles, like atomic nuclei, need more energy to be accelerated close to the speed of light. Some polarization is expected near the edge.

Bubbles and clusters are in equilibrium between expansion and contraction: as bubbles expand in the universe, they keep an outward pressure that counteracts gravitational attraction between clusters. Whenever the combination of bubble pressure and gravity forms a critical massive knot of material, such as a cluster of galaxies with a mass of about one thousand galaxies plus enclosed gas, this same knot will collapse on itself before bouncing back and expanding as a new bubble in a new cycle. So all cosmic material is either part of the expanding kinetic energy of a bubble or tied by gravity to large masses such as stars, galaxies, clusters, and black holes. This way, the cosmic principle of homogeneity and isotropy is respected at large scales, and there is no privileged center in the universe as stated by the Copernican principle. The cosmos is a self-regulated cyclic system of growth and contraction like the cells, in a body with a constant volume, are born and grow to replace dead cells. It may also be compared with the renewed bubbles in a flute of champagne.

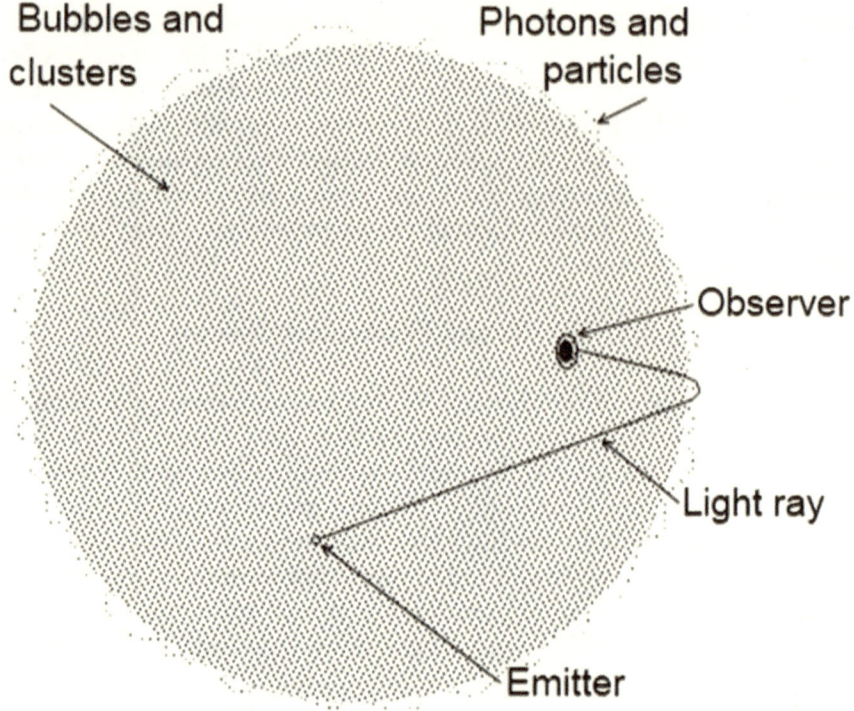

Figure 2.15 Spheroidal multibang cosmos

One of many possible electromagnetic wave or light trajectory between an emitter and an observer is also shown on the figure. We want to emphasize the fact that photons are always deviated when traveling across the cosmos by refractions and lensing effects from the gravitational pull of a multitude of massive bodies and especially by the cosmic mass itself. Moreover, light may be absorbed and reemitted as reflections, refractions, or scatterings. But, in about 90% of the cosmos, light reaches its maximum velocity, c, and is little hampered when it travels across vast, expanding bubble volumes in which extremely low densities and weak fields prevail. So these bubbles may be considered as freer nonreactive corridors and fast tracks for light across the cosmos.

Selected cosmic parameters:			
Physical parameter	Number	Units	Remarks
distance: 1 parsec (pc)=	3.26	light-years (l-y)	1 kiloparsec= 1 kpc
distance: 1 l-y=	9.47E+15	meters (m)	E means exponent in column 2
cosmic density	0.21	Mp/m^3	proton mass/cubic meter
average cosmic radius	1.43E+11	l-y	light-years
average cosmic radius	1.3542E+27	M	Meters
cosmic volume	1.0403E+82	m^3	cubic meters
cosmic mass	2.1846E+81	Mp	proton masses
proton mass	1.67E-27	Kg	Kilogram
cosmic mass	3.65E+54	Kg	
bubbles mass	3.65E+53	kg	10% of total
inter-bubble mass	3.28E+54	kg	90% of total
bubbles volume	9.3624E+81	m^3	90% of total
bubbles mass	2.18E+80	Mp	proton masses
bubble density	0.02333	Mp/m^3	proton mass/cubic meter
neutrino density	1.15E-03	Mp/m^3	1.00E+8 neutrinos/m^3
neutrino mass	2.00E+52	Kg	0.001 to 0.01 of total cosmic mass
electromagnetic density (CMB)	3.00E-04	Mp/m^3	3.3E+8 photons/m^3 and 5E-31 kg/m^3
electromagnetic mass	5.21E+51	Kg	CMB intensity dominated

Table 2.1 Selected bubble and interbubble cosmic parameters

In the first column of table 2.1, we find the physical parameters being evaluated, with the quantitative result in the second column, followed by the corresponding unit in the third column, and comments, as applicable, in the last column. The value of cosmic density is determined from two separate sources: with the baryon density from nucleosynthesis as discussed above and from a physical evaluation of baryonic material in the cosmos. The mass of bubbles is tentatively evaluated as 10% of the total cosmic mass, but it could vary around this value, some authors going as low as 5%. Expressing cosmic-scale densities in term of proton masses per cubic meter is a practical unit since hydrogen (H) is the most basic and most spread element in the cosmos, with a single proton having a mass 1836 times larger than its electron in the neutral configuration.

Neutrino data are still under experimental scrutiny since the mass is only estimated in a bracket between 0.001 and 0.01 of total cosmic mass, suggesting an influence less than 1% of the total. It is evaluated that the cosmic microwave background (CMB), which contributes the most to the electromagnetic energy density of the cosmos, about 30 times more total intensity than the far infrared background (FIR), represents a scant value around 700 times less than the baryonic mass. We have computed a mass energy equivalent for the central CMB frequency of 218 GHz using the Planck formula and found a mass equivalent of 1.6×10^{-39} kg/photon, so we can corroborate the photon to baryon ratio of 1.65×10^9 photons per baryon from CMB as used in figure 2.13.

2.3.6 Friedmann's Equation and Big Bang Model

Einstein's special theory of relativity is only valid for systems at constant velocities when no forces are acting, which means no acceleration and no gravitational field present. The theory predicts the equivalence of mass and energy and also time stretching and space contraction. General relativity, on the other hand, is based on the principle of equivalence between gravitational and inertial mass or in other words between gravitational attraction and acceleration.

In 1922, Friedmann derived this simple equation from the field equation of general relativity to describe different big bang cosmological models of the universe:

$$H^2 = (8\pi G\rho/3) - (kc^2/R^2) + (\Lambda/3) \qquad (2.5),$$

where Hubble's constant is $H=(1/R)(dR/dt)$ and its present-day value is $H_0 = 71$ km/s/Mpc or $H_0 = 2.3 \times 10^{-18}$/s, the gravitational constant $G = 6.67 \times 10^{-11}$ n m^2/kg^2, the critical cosmic matter density $\rho = 5.67$ M_p/m^3 when $k=0$ and $\Lambda=0$, the cosmic curvature $k=+1$, 0 or -1 for the respective models of a close, flat, or open universe, c is the speed of light, R the scale factor as the universe expands, and Λ is the cosmological constant. In this equation, the cosmic mass-energy is represented by density ρ that varies in proportion to the rate of expansion H. The cosmological constant Λ was first introduced by Einstein to counteract the pull of gravity in a model of a static cosmos in equilibrium, but he removed it when the high likelihood of cosmic expansion was introduced by Hubble. Recently, it was reintroduced as dark energy in the big bang model to explain a suggested acceleration of expansion based on supernovae Ia surveys.

Thus, in the big bang model, the critical density at 5.67 M_p/m^3 is far above the value found from the physical count of stars, galaxies, and clouds, which represents an estimate of about 0.21 M_p/m^3 and far from the nucleosynthesis estimate at 0.21 M_p/m^3. As the whole cosmos expands and cools down, the volumetric density should gradually and irremediably be diluted in proportion to the cube of cosmic radius since measured density is insufficient to stop expansion and suggests exclusively an open universe with a curvature k=-1 regardless of other observations. In consequence of the cooling, the universe was denser and hotter in the past up to its birth out of an unexplained singularity about 13.7 billion years ago. The big bang proponents had to invent an early period of rapid expansion and mixing, called inflation, to create a homogeneous and isotropic cosmos and avoid the problem related to the cosmic expanding shell carrying a higher density and surrounding a huge cosmic size internal void.

The average cosmic value of 0.21 M_p/m^3 is used in the multibang model where expansion is expressed by every expanding bubble born from massive knots of mass-energy and where local expansions, corresponding to lambda in Einstein's field equations of general relativity, counteract the pull of gravity everywhere in the cosmos. Also, note that in this model, density is about 10 times lower in the central void of each bubble since mass-energy is concentrated in the expanding shell and contact walls with other expanding bubbles and will be diluted in a proportion close to the square of bubble radius. For example, a bubble with an average radius of 28 Mpc will concentrate three to four times more mass in its shell as compared with its large internal void. Every bubble is gradually cooling after a maxibang, and so is denser and hotter in its past.

Thus, a major difference between the two models, in the formation of extremely low-density voids inside the bubbles across the cosmos, is that in the big bang, material is pulled by the gravity of a few clusters and groups outside of the voids while the whole cosmos expands, like pulling at a few points on the skin of a deflated balloon from the outside and hopelessly hoping to end up with a spherical or polyhedral form as physically observed at large scale. On the other hand, in the multibang model, every bubble is inflated from the inside after a maxibang event and the ensuing expansion of a shell that leaves behind a lower density void, similar to the supernova expansion represented on the front cover of this book.

3. Electromagnetic Waves, Doppler Effect, and Refraction

What is this almost ubiquitous substance without which there would be no astronomical observations and without which astrophysicists could not apply or extrapolate the laws of physics? Of course, it is light and its electromagnetic wave family, which also has a double nature expressed not only as waves but also as photon particles bearing energy proportional to frequency and inversely proportional to wavelength [60].

Class	Wavelength	Frequency	Energy	
Gamma rays	1 pm	300 EHz	1.24 MeV	
X-rays	1 nm	300 PHz	1.24 keV	
Ultraviolet				Visible light
Infrared	1 μm	300 THz	1.24 eV	
Microwave	1 mm	300 GHz	1.24 meV	
Television-FM	1 m	300 MHz	1.24 μeV	
AM radio	1 km	300 kHz	1.24 neV	
EM voice Extemely low	1 Mm	300 Hz	1.24 peV	
Planck's constant	300 Mm	1 Hz	4.14 feV	

Table 3.1 Electromagnetic waves summary

The symbols in Table 3.1 are as follows: m means "meter" when standing alone or following another symbol; Hz means "Hertz" or "cycle per second"; eV means "electron volt." The first symbol preceding m, Hz or eV gives in decreasing order the size in multiples of 1000, except for the last one related to Planck's constant:

E means "exa" and multiplied by	10^{18}
P means "peta" and multiplied by	10^{15}
T means "tera" and multiplied by	10^{12}
G means "giga" and multiplied by	10^{9}
M means "mega" and multiplied by	10^{6}
k means "kilo" and multiplied by	10^{3}
$none$ means "unity" and multiplied by	$10^{0}=1$ (no symbol present)
m means "milli" and multiplied by	10^{-3} (not to be confused with $meter$)
μ means "micro" and multiplied by	10^{-6}
n means "nano" and multiplied by	10^{-9}
p means "pico" and multiplied by	10^{-12}
f means "femto" and multiplied by	10^{-15}
a means "atto" and multiplied by	10^{-18}

But in astronomy and astrophysics, scientists often exceed these values both at the small- and large-scale ends.

The first column of table 3.1 shows electromagnetic waves (EM waves) from gamma rays at the high-energy end to Planck's constant at the lower limit and where energy E is the difference between two quantum states and equal to the product of Planck's constant h by the frequency f:

$$E=hf \qquad (3.1).$$

This is called the Planck formula, in which frequency f is also proportional to the speed v of the wave divided by wavelength λ:

$$f=v/\lambda \qquad (3.2),$$

where v may be replaced by symbol c for the speed of EM waves.

A thin line across the chart indicates the position of visible light, which is only a tiny portion of the electromagnetic waves.

3.1 Doppler Effect in Daily Life

After a review of Doppler's effect related to movements of source and/or receptor, we will look at refraction in a medium and present the prism as an example [59]. Astrophysics would not be the same if the Doppler effect associated with light was not present in the cosmic realm. But as earthly observers, we experience the effects of Doppler shift in our daily lives, mostly with sounds. For example, when a train moves toward us, the apparent sound frequency is higher than emitted frequency; and when it moves away, frequency is lower than emitted frequency.

3.1.1 Doppler Shift Related to Source and Observer

The emitter is the source of the wave disturbance, and the receptor represents the observer in the bottom sketch of figure 3.1 where four complete cycles are represented. When a sound or light wave is emitted by a source, it travels across the surrounding medium- or low-density field while it is apparently stretched or contracted and until it reacts by absorption, refraction, reflection, or scattering.

Note that sound waves are carried by the medium in which they travel, like the wind, for which the velocity of the wind adds up or is subtracted from the velocity of the sound wave depending on direction. Light or EM waves, on the other hand, react with the surrounding medium in relation to its density, but the Michelson-Morley experiment demonstrated that light is not carried by a medium such as ether, so the speed of light is considered as a constant for any moving reference frame such as earth.

Light travels at a maximum official speed of c=299,792,458 m/s in the low-density medium of the solar neighborhood and should travel at a slightly higher speed in an extremely low-density intergalactic medium. In practice, we will use a rounded value of c=300,000,000 m/s in our calculations. In our scale, we are not sensitive to Doppler light shifts on earth because the speed of light is so much larger (about one to 10 million times more) than our familiar speeds, such as a car at 30 m/s and even sound at a speed of only 330 m/s. But in our human scale, Doppler shifts from sound waves transmitted in air as a medium are the most familiar and easy to measure since the car speed at 30 m/s is a substantial fraction or about 9% of the speed of sound at 330 m/s.

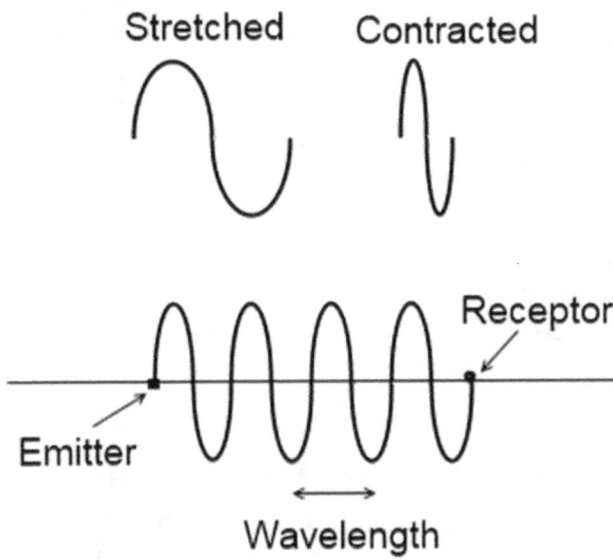

Figure 3.1 Doppler shift related to fixed or moving emitter and receptor

When both emitter and receptor stand still in a homogeneous medium where friction and other energy losses are neglected, then the wave is transmitted with a constant frequency, wavelength, and amplitude, and no Doppler effect is present. If either the emitter or the receptor or both are moving toward each other or away from each other, then Doppler shifts may be recorded. Whenever the net displacement between emitter and receptor brings them closer, then frequency apparently increases, and wavelength is contracted as shown on the upper right of figure 3.1, and this translates into a blueshift if light is involved.

On the contrary, if the displacement brings emitter and receptor apart, then frequency apparently decreases, and wavelength is stretched as shown on the upper left of the figure, and this corresponds to a redshift if light is involved.

3.1.2 Frequency Shift when Receptor Is Moving away

The Doppler effect when the observer or receptor is moving away is illustrated in figure 3.2 using only one cycle in one second (1 Hertz) to visualize more easily the phenomenon. For sound waves, we would multiply the number of Hertz by 20 to 20,000 for the range of human audition.

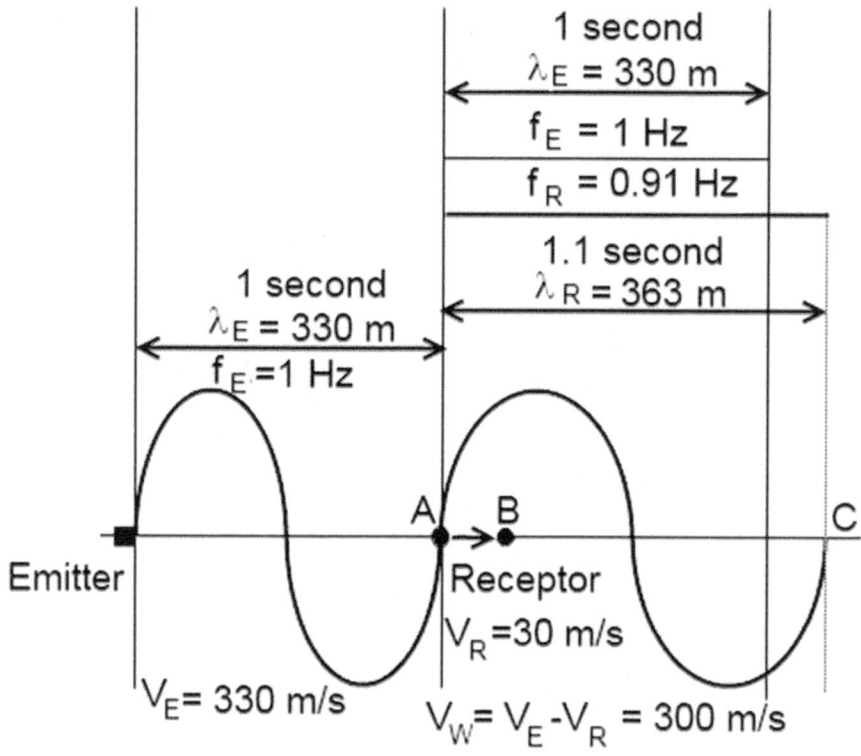

Figure 3.2 Doppler effect when receptor is moving away

The vibration is generated by the emitter and travels in the air as a medium at a speed of 330 m/s and a frequency of 1 Hertz. From equation 3.2, we find a wavelength of 330 m, and since the medium does not change, the speed, frequency, and wavelength of the vibration are the same, second after second, away from the emitter and independently from any observer, as shown after two seconds on the upper right of the figure.

But if after one second an observer is moving from point A to point B at a speed of 30 m/s and away from the source emitting the signal, then the signal has to catch up with him, and he will measure a lower wave speed V_W of 300 m/s for that signal, which is equal to the difference between emitted speed V_E and receiver speed V_R, so we have

$$V_W = V_E - V_R \qquad (3.3).$$

Combining equations 3.2 and 3.3, we find $f_R = V_W/\lambda_E = 0.91 Hz$, which is the observed lower frequency representing less than a complete cycle in one second. But to receive a complete wavelength or a full cycle of 330 m from the emitter, our moving observer needs more time. He receives the signal at a speed of 300 m/s, which means that the time needed is $t = \lambda_E/V_W$ and equal to 1.1 second and corresponds to an observed wavelength of $\lambda_R = \lambda_E t = 363$ m.

Note that the energy of the emitted wave is basically conserved since it is stretched out and received over a longer period for the observer. The apparent shift is related to displacement and proportional to the variation of distance between emitter and receptor. It is positive when distance increases and negative when distance decreases. In astrophysics, the distance variation may also be replaced by space expansion and result with the same effect.

The above results can also be expressed as the difference between emitted and observed frequency and divided by the observed frequency corresponding to a positive frequency shift, $\Delta f/f_R$, called a redshift when light is involved and where we have

$$z = (f_E - f_R)/f_R \qquad (3.4),$$

which may also be expressed as

$$1+z = f_E/f_R \qquad (3.5)$$

or the equivalent positive wavelength shift, $\Delta\lambda/\lambda_E$, or redshift

$$z = (\lambda_R - \lambda_E)/\lambda_E \qquad (3.6),$$

which may also be expressed as

$$1+z = \lambda_R/\lambda_E \qquad (3.7),$$

where $z = 0.1$ is the redshift equivalent in our above example. Then equation 3.5 yields $1.1 f_R = f_E$ and equation 3.7 will similarly yield: $1.1\lambda_E = \lambda_R$ in which the only difference between emitter and receiver is the stretch factor. Contrary to positive shifts, the convention requires that a frequency or wavelength shift, with a negative z value, corresponds to a blueshift.

3.1.3 What Happens when a Balloon Separates the Source and Receptor?

Figure 3.3 represents a wave emitter and a receptor similar to figure 3.1 but with one important difference: a spherical balloon represented in 2-D has been added to the figure. Again, if the emitter and receptor located anywhere across the balloon are not moving, waves are transmitted with constant amplitude, frequency, and wavelength. But if we assume that the sphere is expanding in a homogeneous medium and that densities inside and outside the balloon remain equal or vary with a negligible difference, then as the sphere expands, both emitter and receiver move away from each other, so the distance between them increases, and the relative signal speed V_W decreases. Therefore, the relationships between speed, frequency, and wavelength are equivalent to our example in figure 3.2.

The point is that the signal entering anywhere on the right side of the balloon is stretched while the balloon expands, and this in proportion to the total expansion velocity or to the stretched distance when reaching the observer: it is positively Doppler shifted or redshifted for light or any EM wave. The inverse would apply if the balloon was shrinking instead of expanding; the signal would be negatively Doppler shifted or blueshifted in the case of a light wave or any EM wave. This situation may be transposed to an earthly observer measuring the redshifts of galaxies and clusters located on the other side of our local, northern or southern, expanding and clashing bubble if and only if we respect the fact that light is not traveling in a medium like air or ether but in a low-density field where space is expanding within a bubble and stretching EM waves. Moreover, light coming from two or more bubbles away will be stretched or contracted in proportion with the expansion or contraction of every bubble crossed. But let us be clear: no new space is created within the cosmos.

Big Bang out, Multibang in

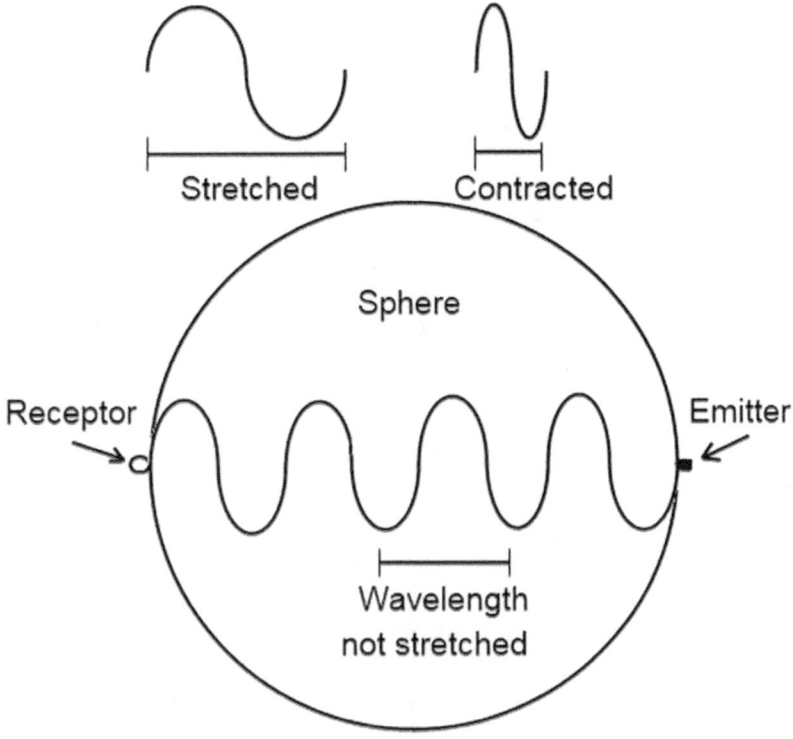

Figure 3.3 Expanding sphere between emitter and receptor

On the top left of the figure, a positive Doppler shift is illustrated as a stretched wavelength that occurs when the balloon expands; and on the top right, a negative Doppler shift corresponding to a decreasing balloon is shown as a contracted wavelength.

3.2 Cosmic Light Redshifts

An illustration of cosmic light redshift is presented on table 3.2 for a small portion of the electromagnetic spectrum, including the visible wavelength window between 380 nm and 740 nm and nearby part of the invisible ultraviolet shorter than 380 nm and infrared wavelength longer than 740 nm. A redshift of z=0.2 is represented in the central column between reference absorption line values at z=0 for the sun in column 2 from the left and redshifted absorption line values from faraway galaxies shown in the fourth column. The interpretation of redshift is questioned by some specialists such as Halton Arp, who suggests the presence of intrinsic redshift not related to distance or expansion [2].

Wavelength nm	Sun	Z=0.2	Redshifted	
300	Ultraviolet	→	Ultraviolet	Invisible
380 nm				
400	Violet / Blue		Violet / Blue	Visible light
500				
600	Green / Yellow / Orange		Green / Yellow / Orange	
700	Red		Red	
740 nm				
800	Infrared		Infrared	Invisible

Table 3.2 Redshift illustrated from ultraviolet to infrared

We notice that redshifting increases with wavelength as we go down the table. For example, an ultraviolet absorption line at 300 nm in the sun is stretched by an amount of 60 nm to 360 nm in our faraway galaxy, but an absorption line, at twice the 300 nm wavelength, in the yellow light of the sun at 600 nm would be stretched twice as much by a length of 120 nm to 720 nm in the faraway galaxy. Therefore, stretching or redshifting is proportional to wavelength [74].

When we observe visible light from farther and farther away objects, we are looking at more and more redshifted objects until we reach the invisible infrared waves which may be observed with infrared instruments. Of course, some of the high energy emitted in the ultraviolet or higher frequencies by faraway galactic clusters or quasars may be redshifted and observed in the visible or at lower frequencies, thus increasing the range of the observable universe. Also, note that in the big bang model, distance increases with redshift up to the limit of 13.7 G l-y, and the age of objects decreases with redshift down to Planck's time. But in the multibang model, distance increases with redshift without any limit within the 143 G l-y cosmic radius, and the age of objects only depend on the age of the bubble from which they are born.

Figure 3.4 Comparison between one and two cycles for redshift z=0 and z=1

This redshift-related stretching is further illustrated in figure 3.4, where the values of z, indicated in the first column, are computed in the fourth column on the right using the wavelength equation 3.6, where subscript o means "observed" and subscript e means "emitted." One- and two-cycle waves are shown with the scale of one unit in the second column and where z=0 at the top and z=1 at the bottom, corresponding to a stretching that doubles the wavelength. The important point to notice is that the longer wavelength (lower frequency=one cycle) is stretched by one unit while the shorter wavelength (higher frequency=two cycles) is only stretched by ½ unit, confirming that redshift is proportional to wavelength.

Now if we compare z=0 with z=1, we realize that energy is conserved since it takes twice the wavelength distance for z=1, which corresponds to twice the time for the frequency. We know from equation 3.1 that E=hf, so for example, the frequency of one cycle per unit time at z=0 will be cut by ½ and take two time units after stretching at z=1 and will have the same energy: E=h1=h2/2. Therefore, light coming from distant galaxies and redshifted by bubble space expansion conserves its energy since the same energy is distributed over a longer stretch corresponding, at constant speed c, to a longer period of time. Then we have the following relationship that applies:

$$E=h(1+z)f \qquad (3.8)$$

or

$$E=h(1+z)c/\lambda \qquad (3.9).$$

Thus, the expansion factor $(1+z)$ redistributes the energy over the stretched EM wave, so in the cosmic microwave background, we should find all the emitted energy from all sources conserved in a stretched form.

3.3 Refraction across a Medium

Electromagnetic waves interact with the transparent media that they cross and this interaction is called refraction and may be expressed as a refraction index n equal to the ratio of the speed c of EM waves in a vacuum to the speed v of EM waves in a medium:

$$n=c/v \qquad (3.10).$$

For example, the index of refraction in air is $n=299{,}792/299{,}704=1.000293$, where c and v are in km/sec, and the index of refraction of zircon is $n=299{,}792/155{,}898=1.923$.

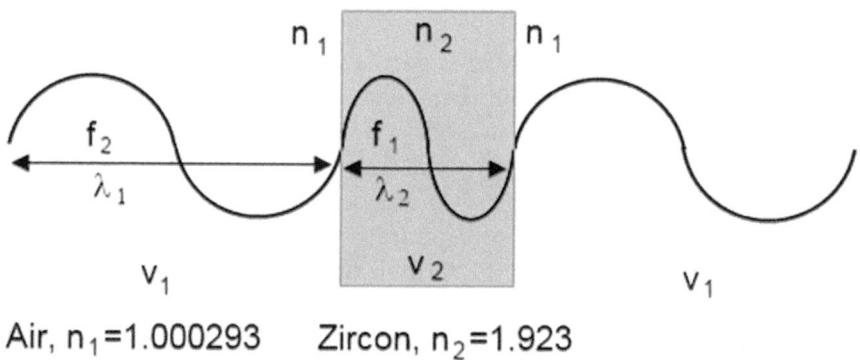

Figure 3.5 Refraction across a denser medium

In figure 3.5, we apply these two refraction indices to an EM wave with an index of refraction n_1 in the air, entering a crystal of zircon with an index of refraction n_2 from the left and then reappearing in the air to the right. Since the index of refraction of zircon is almost twice of that in air and since the speed of

EM waves in zircon is almost half of that in air, we find that the wavelength in air is almost double of that in zircon and establish the following relationships:

$$v_1/v_2 = \lambda_1/\lambda_2 = n_2/n_1 \qquad (3.11),$$

where the frequency in the two media is the same $f_1=f_2=f$. That means velocity and wavelength in a medium are proportional to the index of refraction in the second medium. Thus, we expect to find a variety of refraction phenomena across higher-density clouds of gas and dust and filamentary structures in flat walls or in 3-D galactic and intergalactic space.

One interesting implication from the slower speed of EM waves in most media as compared with c is that neutrinos and neutrons can travel faster than EM waves in dense media since they don't interact like EM waves. It is theoretically possible to find neutrinos across denser clouds very slightly faster than the value of c since the latter could have been underestimated in our solar neighborhood, especially since the speed of light, c, has never been measured in the extremely low-density bubbles of intergalactic space.

3.4   Refraction in a Prism

We have seen in equation 3.11 that wavelength is inversely proportional to the index of refraction. In figure 3.6, we see that the index of refraction n at the two interfaces of a prism varies inversely with wavelength λ: the white light is spread in its different wavelengths at the first interface, and this effect is amplified by the second interface from which we can see that the red component with the longest wavelength has the lowest refraction index since it is the least deflected and takes the shortest path; on the other hand, the blue component has the shortest wavelength and the highest refraction index since it has the greatest deflection and takes the longest path [25].

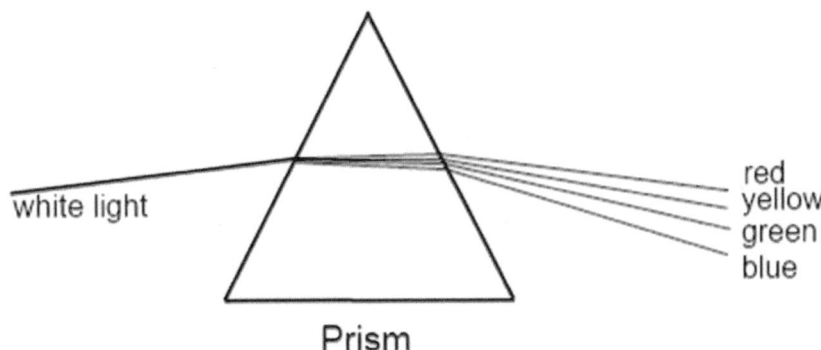

Figure 3.6 Refraction of light across a prism

For example, a glass prism may have a refraction index n=1.535 for the blue wavelength at 380 nm and a refraction index n=1.510 for the red wavelength at 740 nm, which makes a difference Δn=0.025 in the index of refraction and explains the color separation.

Although the expected index of refraction across wedges of clouds and filamentary structures in galactic and intergalactic space is by far smaller than in the example above, it is not null and may affect significantly the mixture of light and other EM waves observed anywhere in the cosmos. This implies that light coming from high z regions and traveling over vast distances across a large number of bubbles may be doped by different wavelengths from a variety of its neighboring EM sources. It also means that this effect may be misinterpreted as gravitational lensing, the latter being the gravitational deviation of EM waves by large masses acting like a lens along the trajectory of photons between source and observer.

4. EXPANSION AND REDSHIFT, LIMITED CONTRACTION AND BLUESHIFT, COSMIC MICROWAVE BACKGROUND (CMB)

After a summary of the bang time sequence for our northern bubble and for the 2dFGRS data slice across part of the cosmos, we will concentrate on bubble expansion and related redshift, followed by light trajectories and multiple bubble expansion and redshift. The link between redshift and distance will also be presented in relation to a q correction factor that will compensate for limited expansion or no expansion and for blueshifting within some cosmic volumes. Finally, the cosmic microwave background will be presented as the blackbody from all higher cosmic EM frequencies from all sources that have been freely redshifted to the CMB frequencies and boosted to the peak of the CMB by the Sunyaev-Zel'dovich effect over the entire cosmos for trillions of years.

4.1 Observation of Data Slices across the Cosmos

In the past 10 years, a substantial amount of redshift space data has been gradually published to cover large volumes and slice volumes of the nearby cosmos [44]. The two most representative include the two-degree Field Galaxy Redshift Survey, which is second in size, referred to as the 2dFGRS and presented in figure 4.1, and the most voluminous and more recent ongoing Sloan Digital Sky Survey (SDSS) that will register data for over 100 million galactic objects.

The 2dFGRS was completed in 2003 and measured the photometry of about 380,000 objects and, among these, the spectrum of about 250,000 galaxies from two separate right ascension bands of 75°, with respective north and south declination of 7.5° and 15° in the direction of the galactic poles. The four-meter Anglo-Australian telescope was used for the survey in which two square degree pictures were recorded per exposure.

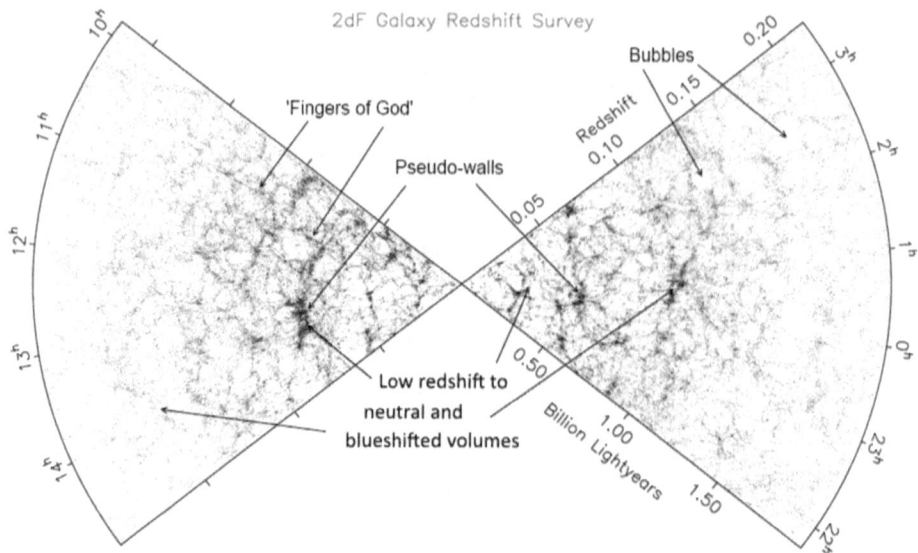

Figure 4.1 Two-degree Field Galaxy Redshift Survey

We are located at the center of this picture, where galactic north is to the left and where each tiny dot in the figure represents a galaxy. The scale is about 0.92 billion light-years for the redshift z=0.1. We observe on this 2-D display of the 3-D slice a multitude of more or less flocculent bubble-like features of various sizes and shapes, mainly polygonal, circular, or elliptical, which have been compared to soap bubbles, with galaxies floating all around the surface of the bubbles. The flocculent aspect is partly due to the piling up of 3-D data points on a flat surface. Pseudo-wall artefacts appear at about the same distance in north and south directions, and they occur where maxima of galactic data points and luminosity input are reached. Remember that we are not at a privileged location in the cosmos (Copernican principle), although large-density fluctuations are recorded. The quantity of information decreases toward the edge of these slices in proportion with distance and some spherical divergence. Many so-called fingers of God pointing in the direction of the observer are visible; these are generally associated with a line of sight spread of redshift values measured from clusters where galaxies are falling in toward the center of the cluster as explained in figure 4.2. Furthermore, the pseudo-walls and narrow slits perpendicular to the line of sight are partly the location of low redshift, neutral shift, or blueshifted volumes, where redshift is not equal to distance. They represent slowly inflating, neutral, or locally blueshifting bubbles. They are the locus of very low or even negative Hubble rates. This explains the presence of two side-by-side walls in many areas of figure 4.1. The only way to reconstitute the true time-distance from redshifts is to reinflate each separate bubble according to its particular history and normalize the peculiar rates of expansion. Otherwise, astrophysicists and cosmologists will remain glued in quagmire models.

Figure 4.2 Real and observed cluster

The real cluster shown has one or more central massive galaxies surrounded by a large number of infalling galaxies at various velocities. Since the central galaxies near *b* have negligible velocities within the cluster and that galaxies near *d* and *e*, perpendicular to the observer, are not Doppler shifted by peculiar velocities, then these galaxies along the line *d*, *b*, and *e* are observed in their real position. But for all the other galaxies, observation is skewed by light shifting: galaxies near *a* will be observed as blueshifted toward the observer across the central part, and inversely, the *c* galaxies will be redshifted away from the observer and on the other side of the central galaxies, thus forming the elongated shape in the observer's line of sight, also called fingers of God.

This applies to clusters observed between neighboring bubbles. For example, in a real cluster whose center is located at a redshift z=0.01, a galaxy at position *c*, falling in toward the center at a speed of 750 km/s, would have a redshift z=0.0125, larger than the cluster's redshift, which indicates a redshift-based distance located on the far side of the cluster at *a*. Inversely, a galaxy on the far side at *a* would have a redshift z=0.0075, smaller than the cluster's redshift, which indicates a redshift-based distance located closer to the observer at *c*.

The 2dFGRS north and south displays are projections in a plane of data points getting thicker and thicker with distance since relatively small angles were used. A small angle slice across bubbles of various sizes and shapes and their projection on a plane are presented in figure 4.3 where redshift values are indicated on the top.

Figure 4.3 Small angle slice and projection on the plane

The width of the shell from the spheroid crossing the slice is projected on a line with black marks, which are in turn projected on a plane at the bottom. We note that the small angle includes more and more material as it becomes wider to the right, so the same-size bubble may look quite different depending on the location and on the portion crossing the slice. For example, the third bubble on the left is only grazing the slice, but it represents many data points.

What is the average part of the bubbles that small angle slices will cross? To answer this, consider figure 4.4 below.

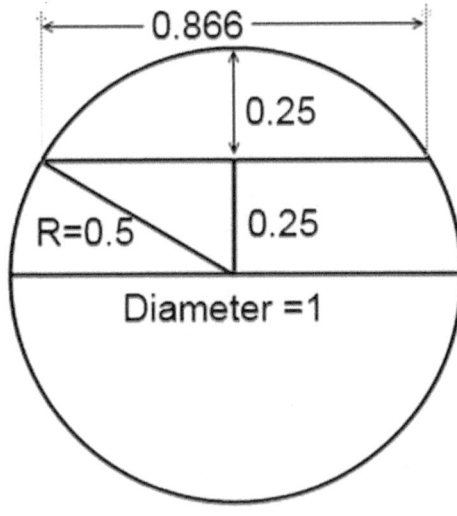

Figure 4.4 Average spheroidal bubble slice crossing

Since we are dealing with a thin slice crossing a large number of spheroidal bubbles, we realize that the average width of the slice is neither the maximum of one in the center or the minima of 0 at the top or bottom, but somewhere in between. We can see that the average width is halfway between the top or bottom and the center, where radius 0.5 is divided by two to yield a distance of 0.25 from the center; thus, from the Pythagorean theorem, the average crossing width is 0.866, or about 87%, for a diameter of one unit. This average crossing at 87% of bubble diameters implies an important correction to measurements of spheroidal or ellipsoidal bubble diameters from surveys such as the 2dFGRS. In other words, the surveys are underestimating the average radius of bubbles, and the corrected radius is $R_{corr} = R_o / 0.866$, where R_o means "observed radius." This correction would not apply to SDSS data if the larger 3-D volume is used instead of small angle slices.

4.1.1 Expansion across Bubbles

Clusters are located around bubbles as sketched in figure 4.5. As shown earlier, there may be as many as six primary hexahedral and eight secondary tetrahedral clusters in contact with each bubble for an average count of one hexahedral and two tetrahedral clusters per bubble.

Figure 4.5 Expanding bubble and clusters

For example, if a bubble expands at a rate of 4,400 km/s across its diameter, then galaxies and clusters located on the expanding shell are moving away from one another, just like dots on an inflating balloon: the farther the dots, the faster they move away from each other. Note that this is a similar situation as the positive Doppler shift observed and defined with figure 3.3, where an expanding balloon was separating the emitter and the receptor. It is the space within the bubble that is expanding and producing the redshift, but no new space is created since the expansion is within the cosmos. So it may be qualified as local cosmological redshift directly associated with all the expanding bubbles that are interspersed with neutral nonexpanding bubbles. The emitted light coming from the left of the bubble is subject to the like of a partial track stand as it moves and stretches across the expanding bubble. For an observer located faraway to the right, this local expansion corresponds to an additional redshift Δz proportional to the bubble expansion velocity v divided by c, the speed of light in vacuum:

$$\Delta z = v/c \qquad (4.1).$$

So $\Delta z = 0.015$ between the far wall and the near wall of the shell. This light shifting across the bubble decreases gradually from the center to the top and bottom, and thus, the shape of the bubble is roughly preserved for the observer, although as seen in the line of sight, a large portion of the top and bottom may be dampened by extinction from dust or dense gas and crowding of galaxies. Therefore, in our model, each expanding bubble apparently stretches EM waves and contributes to the total z in proportion to its own expansion:

$$z = v_1/c + v_2/c + \ldots v_n/c \qquad (4.2),$$

where $v_1, v_2 \ldots v_n$ are the expansion velocities of bubble 1, bubble 2, etc. This may be expressed as

$$z = \sum_{i=1}^{n} V_i/c \qquad (4.3)$$

or as the general relationship

$$z = nv/c \qquad (4.4),$$

which represents, over large cosmic distances, the sum of all the redshifts between light source and observer with v being the average bubble expansion velocity and n the number of bubbles. The Hubble law written with the Hubble constant H_o is

$$H_o = v/d \qquad (4.5)$$

in the big bang model, where H_o is the Hubble constant representing the cosmic rate of expansion in km/s/Mpc at which galaxies are moving away from the observer, v is the velocity in km/s at which the source is moving away, and d is the distance to the source in Mpc. It is only valid as a single bubble rate of expansion in the multibang model and must be modified when more than one bubble is present to read as

$$H_o = nv/d \qquad (4.6)$$

with a different meaning, where H_o is now the average bubble rate of expansion in km/s/Mpc and may be used as a constant over a large distance, n represents the number of bubbles crossed, v is the average expansion velocity in km/s across the bubbles, and d corresponds to the distance between source and observer in Mpc. Replacing the value of v from equation 4.4, we obtain

$$H_o = cz/d \qquad (4.7),$$

where our universal Hubble constant is proportional to the product of the speed of light times the redshift and to the inverse of the distance Since H_o and c are constant values, then in the multibang model, we have distance d increasing monotonically as redshift z increases. This is not the case for large distances in the big bang model where H varies with redshift z.

Cosmic Bubble Parameters			
Physical Parameter	Number	Units	Remarks
number of bubbles	3.4924E+09	Bubbles	for average size
average radius	28	megaparsec (Mpc)	1 Mpc=1 million parsec
bubble density	0.02334	Mp/m^3	proton mass/cubic meter
average radius	9.10E+07	l-y	light-years
average radius	8.6177E+23	M	
average volume	2.6808E+72	m^3	cubic meters
average bubble mass	6.2570E+70	Mp	proton masses
proton mass	1.67E-27	Kg	
average bubble mass	1.04E+44	Kg	

Table 4.1 Cosmic bubble number and parameters

In columns 1 to 4, we find the physical parameters, a number and associated units, plus some remarks. Since bubbles grow from invisible when they are nascent to an average of 28 Mpc radius at maturity, we present only average values in table 4.1. Bubble density is about 10 times less than the cosmic average. The average mass of a single bubble is 1.04×10^{44} kilograms or about three times less than the mass of a single cluster. It includes the mass of the tenuous gas present at the center, where most of the volume is found, but it also includes gas, filamentary structures, and galaxies present in the fossil walls swept by the expanding bubble.

4.1.2 Maxibangs and Expansion

Observing all these bubbles from different surveys with their cohort of galaxies and clusters riding on their back, we are justified to ask, what happens if we contract these bubbles backward? To what origin in space and time is every individual bubble pointing? Or how old are they? We will see that they are all born at separate times somewhere in the cosmos and that the large spread of ages is associated with permanent and dynamic renewal cycles.

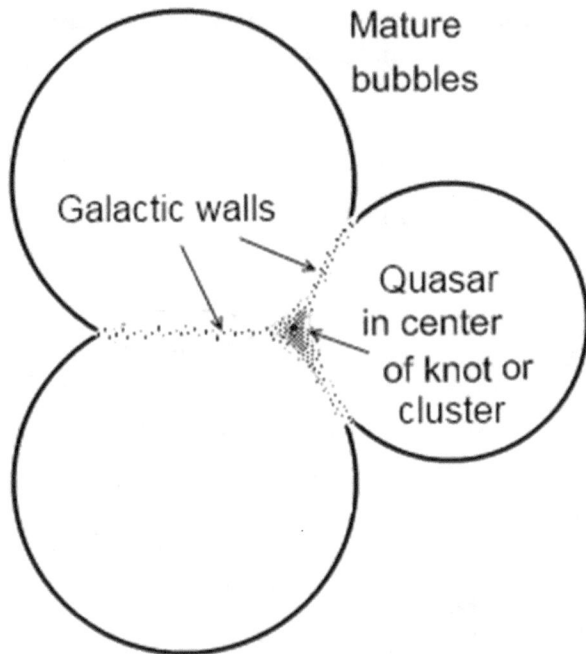

Figure 4.6 Mature bubbles with quasar in center

Little is known about the formation, life, and disappearance of quasars except that they are faraway, often associated with powerful double and opposite jets of material exhibiting a large spectrum that, depending on orientation, may cover the EM range from gamma rays to radio and sometimes output energy comparable to hundreds of galaxies. We put forward the hypothesis that ultimate black holes, most likely quasars are part of the final stage in the contraction of clusters of galaxies by gravity before collapse and rebound as a bubble.

A quasar forming at the intersection of four to six bubbles (only three drawn) is shown in figure 4.6. Mature bubbles have radiuses with an average of 28 Mpc (91 million light-years) and are separated by mostly flat walls composed of groups of galaxies in the process of being squeezed out and pulled toward the closest cluster. At the center of the cluster, a giant galaxy with an active galactic nucleus, or AGN, exhibiting a larger spectrum of EM frequencies is formed by accretion of smaller galaxies. The contraction process continues until a quasar is formed and exhibits its range of characteristics. After an average life of about 350 million years, quasars and their associates reach their final destiny in a gravity crunch from which they bounce back in a maxibang explosion as shown in figure 4.7.

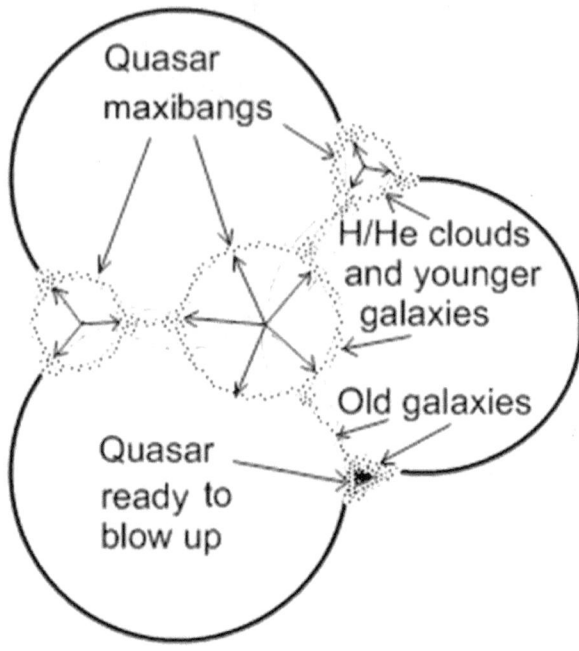

Figure 4.7 Maxibangs from quasars and expansion

Three quasar bangs are sketched in expansion, with shells of hydrogen and helium (H/He) gas forming the growing bubbles, which are gradually cooling down to temperatures where a new cycle of stars and galaxies will start to form. Older galaxies are also shown to indicate that older wall material will mix with the newborn and younger galaxies, so we may find local metal contamination in an otherwise pristine cradle. Also, a quasar surrounded with old merging galaxies is shown ready to blow up. But as the bubbles expand, they fill gradually a larger volume and will merge with their closest neighbors as depicted in figure 4.8. Streaming motion will be observed between neighboring bubbles.

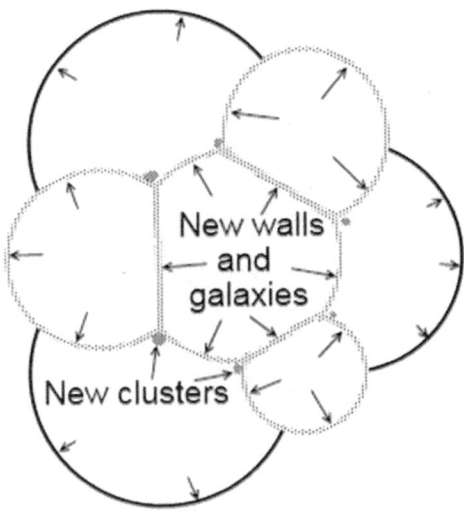

Figure 4.8 New walls, galaxies, and clusters

When two bubbles meet, they form a common dynamic wall that is either flat or curved inward of the larger bubble. Pressure will squeeze the newly twirled galaxies and star clusters toward the edges of the walls to form the next generation of galactic clusters [31]. Note that new bubbles are growing into the volume of older bubbles that are gradually destroyed although they may still be partly expanding at a slower pace in a different direction.

Figure 4.9 Time, velocity, and H values for bubble expansion

Bubble expansion is expressed in terms of age, velocity, and Hubble rate H in figure 4.9. The horizontal scale represents bubble radius from 1 to 50 Mpc which includes most of the cosmic volume. All radius values have been corrected using h=0.71 where h=H_o/100 and where H_o means "average observed Hubble constant" and is equal to 71 km/sec/Mpc from most recent surveys. About 85% of bubble volume is comprised between bubble radiuses of 21 and 42 Mpc corresponding to bubble ages between 5.4 and 15.4 billion years with a bubble expansion velocity range from 3,400 to 5,230 km/sec and a Hubble expansion value H from 40 to 125. The curves on the figure are computed from a normalized escape velocity and corrected for the kinetic energy losses from the average bubble density material that is encountered and added to the expanding shell while slowing down the expanding mass and shedding an amount of material to the growing low-density medium inside the bubble.

Figure 4.10 Void galaxies in expanding bubble

Larger clusters of galaxies are mainly formed in hexahedrons at the confluence of 12 walls and eight triangular sections, and smaller clusters are formed at the confluence of the six walls of four bubbles by the pressure of expanding bubbles combined to gravitational attraction between galaxies. A three-wall schematized version is shown in figure 4.10, where the main feature is the central bubble expanding into four (only three shown in 2-D) larger and older bubbles. Gas, dust, and lower-density material from the six walls between the older bubbles are swept away by the younger and more dynamic expanding shell, but some of the larger galaxies may be preserved in the process to form what are called low-density medium or void galaxies [46]. The latter galaxies are accelerated away by the expanding shell and may retain a fair amount of new gas from the passing shell and become active star-forming galaxies with old cores. Remember that 10% of the cosmic mass is found in the bubbles, which represent 90% of the cosmic volume but only about 5% of the galaxies, mostly as field galaxies located within the internal bubble voids [49].

Fossil walls, filamentary structures, and galaxies within the bubbles:			
Physical parameter	Number	Units	Remarks
number of walls	2.10E+10	walls	6 per bubble
mass of walls	1.64E+53	kg	5% of galaxies
volume of walls	6.5537E+79	m^3	cubic meters
mass of walls	9.83E+79	Mp	proton masses
density of walls	1.50	Mp/m^3	proton mass/cubic meter
Residual extremely low-density bubbles excluding fossil walls:			
residual mass	2.01E+53	kg	in lowest cosmic density
residual volume	9.3624E+81	m^3	cubic meters
residual mass	1.20E+80	Mp	proton masses
residual density	0.01283	Mp/m^3	proton mass/cubic meter

Table 4.2 Fossil walls and extremely low density within bubbles

Physical parameters and associated values of fossil walls and the extremely low-density region within bubbles are shown in columns 1 to 3 of table 4.2, with comments in column 4. As explained for the cosmic geometry, we expect an average of six fossil walls per bubble, and within these walls, we should find about 5% of the galaxies on a cosmic scale. The volumetric density of these walls (1.50 Mp/m^3) is about one hundred times larger than that found in the most desolate voids (0.01283 Mp/m^3), encompassing more than 99% of cosmic bubble volume. The term *fossil wall* means any residual wall material left over when a bubble has ceased to expand and is incorporated by a new bubble.

4.1.3 Light Faster across Bubbles

Void galaxies are reproduced in figure 4.11 in a blowup from the preceding figure. The expanding shell of a bubble leaves behind a low-density vacuum on its track; about 10% of cosmic density will fill this vacuum on average [81].

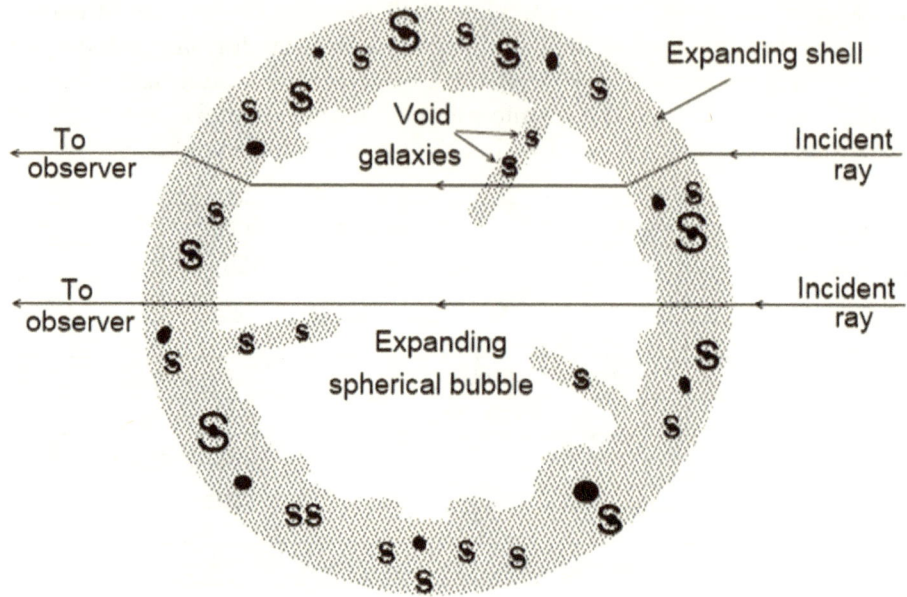

Figure 4.11 Light rays crossing an expanding shell

The speed of light, c, may vary slightly over the vast expanses of the cosmos. When an incident electromagnetic (EM) wave or light ray crosses the center of an expanding spherical bubble, then the maximum EM speed c is reached since it is crossing the lowest cosmic density and since it is located the farthest away from any gravity attractor or magnetic field. If we could measure c in such an environment, we would find the most accurate and fastest value of constant c. Incident rays crossing the expanding shell are refracted and thus slightly slowed down across the thickness of the shell, both at the entrance and at the exit, as shown on the upper part of the figure.

Light coming to us from faraway quasars will cross a large number of expanding bubbles where it is attenuated by absorption in the process. Most of the absorption in the Lyman-alpha forest of quasars is the result of EM waves crossing the shells in the central portion of the expanding bubbles, away from bubble edges where extinction from thick gas and dust may be present and completely obliterate the Lyman-alpha forest and other EM signals. In fact, Lyman-alpha forests of very far quasars will disappear because they are completely attenuated, see figure 5.19.

4.1.4 Light Crossing Time and Expansion

For a better understanding of light crossing an expanding bubble, we will look at the example illustrated in figure 4.12.

Figure 4.12 Light crossing time and expansion

The bubble has a diameter of 60 Mpc (200 M l-y) with an expansion velocity of 4,260 km/sec corresponding to a redshift of z=0.014, since z=nv/c per equation 4.4, and a space expansion rate H_o=71 km/sec/Mpc. This is a radius equivalent to 9.47×10^{20} km with a radial expansion velocity of 2,130 km/sec.

Since light is traveling across the center at speed c of about 300,000 km/sec, it would take 200 M l-y to cross if the bubble was not expanding. But since the bubble is expanding radially at 2,130 km/sec, light will take 201.42 M l-y for its complete wall-to-wall trip; and as shown by the arrows, the bubble will increase its radius by an additional 0.44 Mpc (1.42 M l-y) during that time. From these values, we observe for the spherical model that the ratio of expansion to the original diameter $\Delta d/d=0.014$ is equal to redshift z, so we have a new relationship:

$$z = \Delta d/d \qquad (4.8),$$

which may also be expressed as a ratio of the increase in radius over the original radius for a single bubble:

$$z = \Delta r/r \qquad (4.9).$$

Bubble diameter will increase more and more slowly in proportion to the Hubble rate of expansion H, and it will gradually take more time for light to cross a bubble, and z will increase proportionally.

These distance ratios remind us of the scale factor between the present and the past in the big bang model and related to z by the FLRW (Friedmann-Lemaître-Robertson-Walker) equation $1+z = R_{now}/R_{then}$ where R is the scale factor increasing monotonically as time passes. Thus, the mathematics of the FLRW metric would have to be adapted to our separate and specific expanding bubbles where the rate of expansion is variable instead of the overall big bang cosmic expansion model.

The age of the bubble $t=14$ billion years corresponds to the time taken to cross a radius of 30 Mpc at an average of 2,130 km/sec. Only the first and most significant term is presented here for distances since the second term and followers would add a negligible amount of less than 1.2% to the expansion. It is important to note that because of expansion, light entering at the inner circle on the figure at A would exit at the outer circle at B. Also, note that the above is just an average geometric example that is neglecting the contact and interaction with the neighboring bubbles, and only a numerical example.

4.1.5 Average Light Trajectory across Cosmic Bubbles

For a generally spherical model of renewed bubbles across the cosmos, one question is, where is light crossing a sphere on average? Figure 4.13 will give us a first answer.

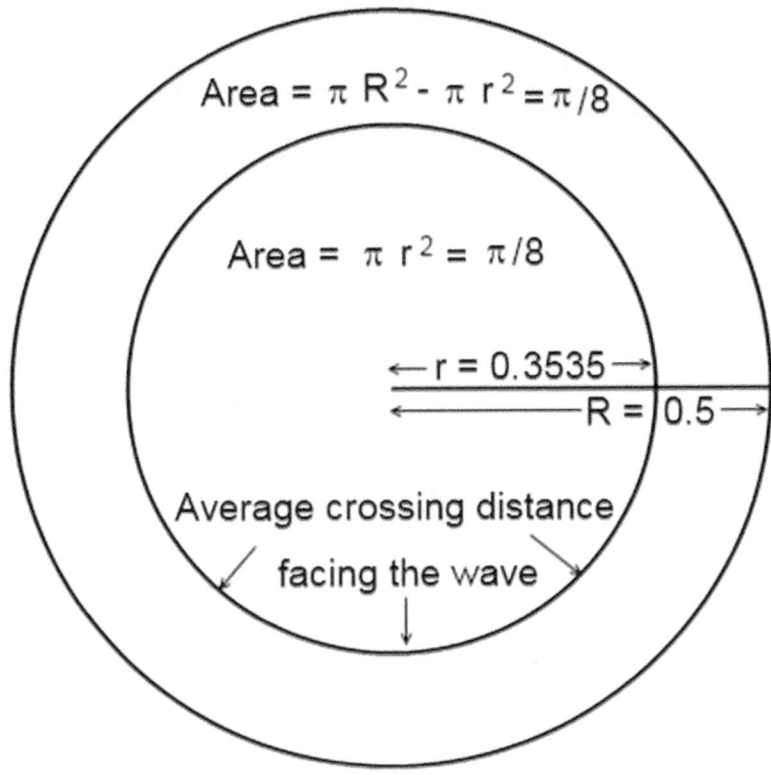

Figure 4.13 Average crossing distance facing the wave

Let us position ourselves facing the EM waves coming at us across any spherical bubble. All that we can see from the sphere is equivalent to a flat circle in 2-D as shown on the figure. From the geometry, we know that the average crossing distance will be located around a circle whose surface πr^2 is equal to the surface ($\pi R^2 - \pi r^2$) of the ring formed between that circle and the edge of the sphere, which implies that $2r^2 = R^2$. In our example, the larger radius, R, is equal to 0.5; and the inner radius is 0.3535, yielding both surfaces with areas of $\pi/8$ and a ratio of r/R equal to 0.7071, which is also the cosine of 45°. Therefore, the average crossing distance is at 0.7071 of the spherical radius, as also shown on the side view of figure 4.14.

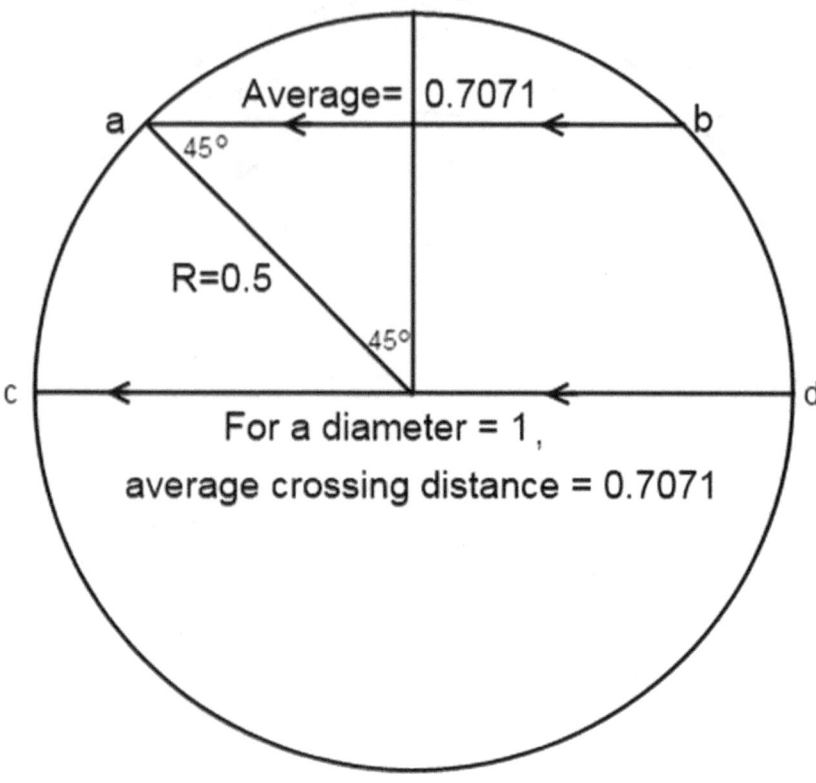

Figure 4.14 Light average crossing distance in a sphere

The side view shows a spherical model with a diameter of one unit between c and d and also the 45° angle associated to the sine or cosine equal to 0.7071. So on average, light crossing a large number of bubbles will enter the sphere at point *b* and exit at point *a* and thus will only travel about 71% of the diametrical distance. If light was always crossing through the diameters of the expanding bubbles, then the rate of expansion would be 100% instead of the average of 71% if we neglect bubble walls in a first approximation.

4.1.6 Local Bubble Expansion

With this average crossing distance in mind and remembering that we are located in a wall between two bubbles, we will concentrate on the best known of the two: the northern one. The center of our local northern bubble is obscured by the plane of our Milky Way galaxy because most galaxies are formed by eddies perpendicular to the wall between two bubbles, and the plane of the galaxies are oriented toward the center of the birth-giving bubbles. We will have a look at the expansion of our local northern bubble and its relationship with surrounding clusters [26] [66].

Figure 4.15 Local bubble expansion with base at cluster A3526

In the absence of a valid model for our local cosmic bubble (not to be confused with much-smaller-scale local bubbles in our galactic molecular clouds near the sun) and the 3-D distribution of galactic clusters surrounding it, we will use a spherical model where clusters are located at their average redshift distance on the surface of the sphere regardless of coordinates but respecting the angles from the observer. We consider light as crossing the sphere at an angle and exiting at the observer station on the left. Angles are viewed by the observer between the center and any point on the surface of the sphere, and they vary between 0° at the surface through the center and 90° at the surface near the observer. They are displayed in steps of 15° with the cosine of the corresponding angle, which gives the expansion ratio between one across the center and zero at the observer. We find our average crossing ratio of 0.707 at 45° as expected, and we are using the Centaurus cluster Abell 3526 or A3526 as a base to locate other clusters.

Cluster A3565 from the Centaurus supercluster is found at about 40°, and Hydra cluster A1060 is located at about 35° from the observer, but in a different direction. Also from the Centaurus supercluster, we find A3574 at about cos 0°, which is the maximum distance of 61.2 Mpc on the sphere, assuming that all these clusters belong to the same bubble. Other clusters, like the Coma cluster and A1367, are located on a separate bubble adjacent to our local bubble and likely separated by a common wall in between [36].

The nearby Virgo cluster likely represents a tetrahedral cluster at the junction of four bubbles, while the larger Coma cluster could be a hexahedral cluster at the junction of six bubbles.

1	2	3	4	5	6	7
For angles from 0° to 90°, every 15°, with cluster A3526 used as a base at 45°						
with z=0.0102 and distance d=43.3						
Angle θ in degrees	Cos θ expansion ratio	z= (cosθ*zbase)/cosθbase	V=cz in km/sec	d in Mpc= (cosθ*dbase)/cosθbase	H_o = V/d rate in km/sec/Mpc	t=d/V time in G years
0	1.0000	0.0144	4,328	61.24	70.7	13.8
15	0.9659	0.0139	4,180	59.15	70.7	13.8
30	0.8660	0.0125	3,748	53.03	70.7	13.8
45	0.7071	0.0102	3,060	43.30	70.7	13.8
60	0.5000	0.0072	2,164	30.62	70.7	13.8
75	0.2588	0.0037	1,120	15.85	70.7	13.8
74.3	0.2706	0.0039	1,171	16.57	707	13.8
77.2	0.2215	0.0032	959	13.57	70.7	13.8
78.4	0.2011	0.0029	870	12.31	70.7	13.8

Table 4.3 Parameters used and computed for our local bubble

Parameters used to compute values for our local bubble in figure 4.15 are displayed in table 4.3. The first column gives the angles in degrees, with the last three values representing, from top to bottom, three different interpretations of the Virgo cluster by de Vaucouleurs, Sandage, and Arp. The second column is the expansion ratio corresponding to the cosine of the angle and decreasing from the center to 90°, not to be confused with the Hubble rate of expansion *H*. The third column presents the average redshift z=0.0102 at 45°, which is the basic observation by astrophysicists from which almost everything else is computed and extrapolated. The velocity V=ncz in kilometers per second is tabulated in the fourth column with n=1 and distance *d* in Mpc in the fifth column with values decreasing from the center to 90°, just like velocities and distances between polka dots on an inflating balloon would increase. We find in the sixth column H=V/d, the Hubble rate of expansion in kilometers per second per megaparsec, which is a constant value within each spherical bubble but varies from bubble to bubble; for a larger bubble, we would find a smaller Hubble value, H<71, and for a smaller bubble, we would find a large Hubble value, H>71. Remember that Ho is an average based on a very large number of bubbles, and it may largely deviate from the average in any bubble. Finally, in the seventh column, the age of our bubble, t=d/V, is shown as 13.8 billion years.

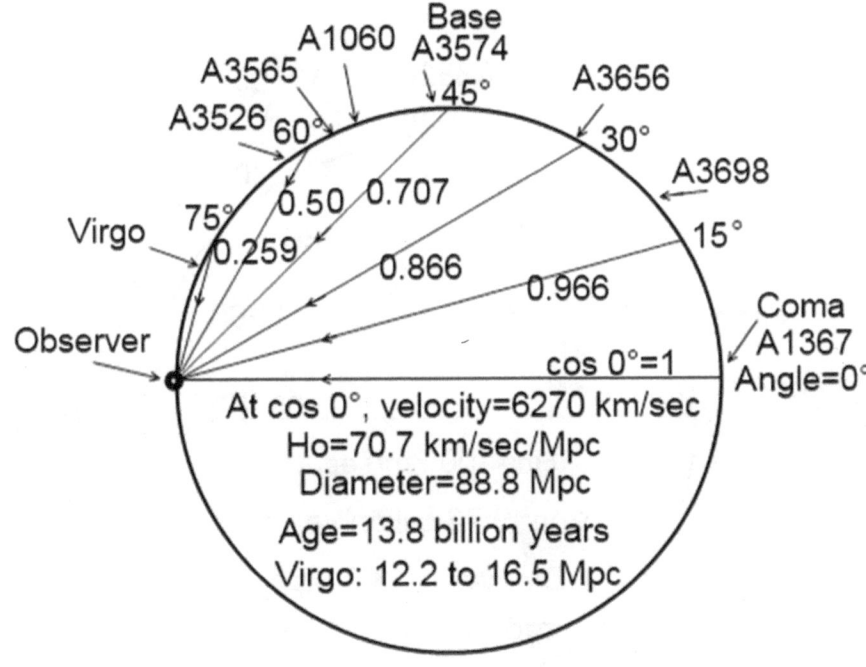

Figure 4.16 Local bubble expansion with base at cluster A3574

If we replace the base for parameter computations with cluster A3574 at 45° instead of cluster A3526 in table 4.3, we obtain the local bubble expansion model in figure 4.16. Since cluster A3574 from the Centaurus supercluster has a larger z value equal to 0.0148 corresponding to velocity V=4,440 km/sec and a distance d=62.8 Mpc, then the maximum at cos 0° is a velocity V=6,270 km/sec and a distance d=88.8 Mpc. All the other parameters, like H, age t, and the three Virgo distances, are unchanged. But the result is a local bubble that does not correspond to some published evaluations attributing a diameter of about 60–62 Mpc to our bubble. So if the larger model were real, then clusters A3698 at 20° and cluster A3656 at 30° from the Pavo-Indus supercluster could be projected on the sphere. Also, the Coma cluster and one of its larger neighbors, A1367, north of the Milky Way, would fall near the maximum diametrical distance of 88.8 Mpc. But between the observer and the Coma cluster, we find a large number of galactic groups forming a wall and suggesting the presence of a separate bubble near the Coma cluster. As mentioned, the large Coma cluster is likely a hexahedral cluster at the junction of six bubbles. It should evolve into a quasar and the next maxibang event in our neighborhood.

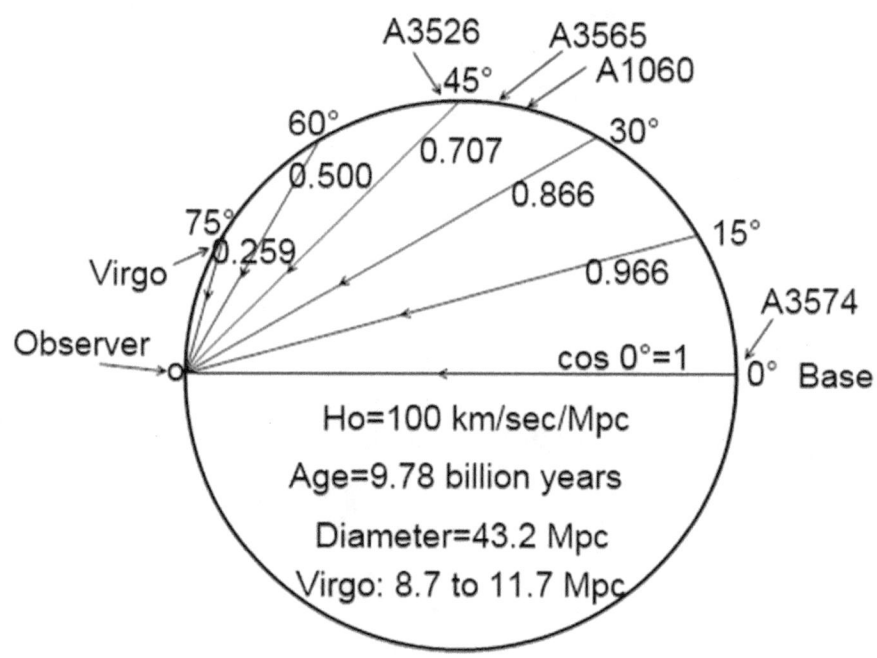

Figure 4.17 Local bubble expansion with base at cluster A3574 at 0°

The local bubble expansion model shown in figure 4.17 has also a difference in the base for computations from that in table 4.3. This time, a position near cluster A3574 at 0° from the Centaurus supercluster is used with a redshift z=0.0144 corresponding to a velocity of 4,320 km/sec as in figure 4.15. This yields a diameter d=43.2 Mpc and a Hubble rate H=100 km/sec/Mpc equivalent to an expansion time t=9.78 billion years. The Virgo distances are between 8.7 and 11.7 Mpc, which is too low by most models based on standard candles like Cepheids and supernovae. This model must be rejected because distances are too small, age is too young, and the rate of expansion is too high.

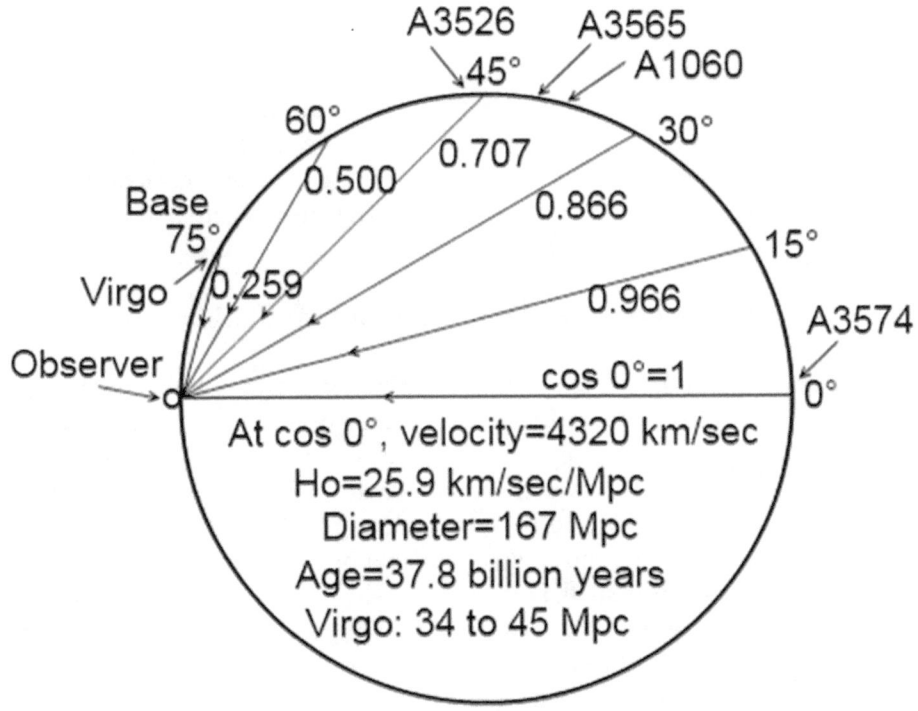

Figure 4.18 Local bubble expansion with base at Virgo cluster

The last comparison model for our local bubble is presented in figure 4.18 using a base at the Virgo cluster near 75° with an expansion ratio of 0.259. This base involves a redshift z=0.00373 corresponding to a velocity V=1,120 km/sec and thus a distance d=43.24 Mpc. The Hubble rate of expansion is only H=25.9 km/sec/Mpc in this conjecture, which yields a large diametrical distance d=167 Mpc and an age of 37.8 billion years for the bubble. These results do not correspond to any known values relative to our local bubble; the age and diameter are way too large and the Hubble rate much too low. The Virgo cluster would be too far by more than a factor of two. Therefore, this model is to be rejected, and the first model in figure 4.15 seems to be the most reliable and closer to known calibrated values. The great advantage of this type of expanding bubble modeling is that it takes care of the very significant expansion ratio in our own and neighboring bubbles, which is ignored in other paradigms.

4.2 Expansion and Contraction between Bubbles

After this presentation of our local universe made of an expanding void and surrounding clusters and galaxies, we will examine some aspects of the interaction between two and more bubbles expanding farther away than in the proximal cosmos. We will see how star clusters, galaxies, and galactic clusters are likely to form in a dynamic process that repeats itself again and again with each bubble expansion and clashing contact.

4.2.1 Dynamic Wall between Two Bubbles

The dynamics resulting from the contact between two bubbles is sketched in figure 4.19 with some of its attributes. When a bubble expands and reaches another expanding bubble, the expanding shells have left behind a void with a density about 10 times lower than the average cosmic density [65]. This low-density medium is made up, in a proportion of about 12.5 hydrogen atoms to one helium atom, from ionized hydrogen and helium atoms drawn from the backside of the expanding shell. So a small density difference should be present between the front wave and the back of the moving shell, creating some density turbulence within the shell. This relatively small turbulence is amplified by the presence of up to 12 old vestigial walls and will produce eddies and twirls all around and within the shell, where cooling and locally increased densities will induce the formation of early stars and star clusters.

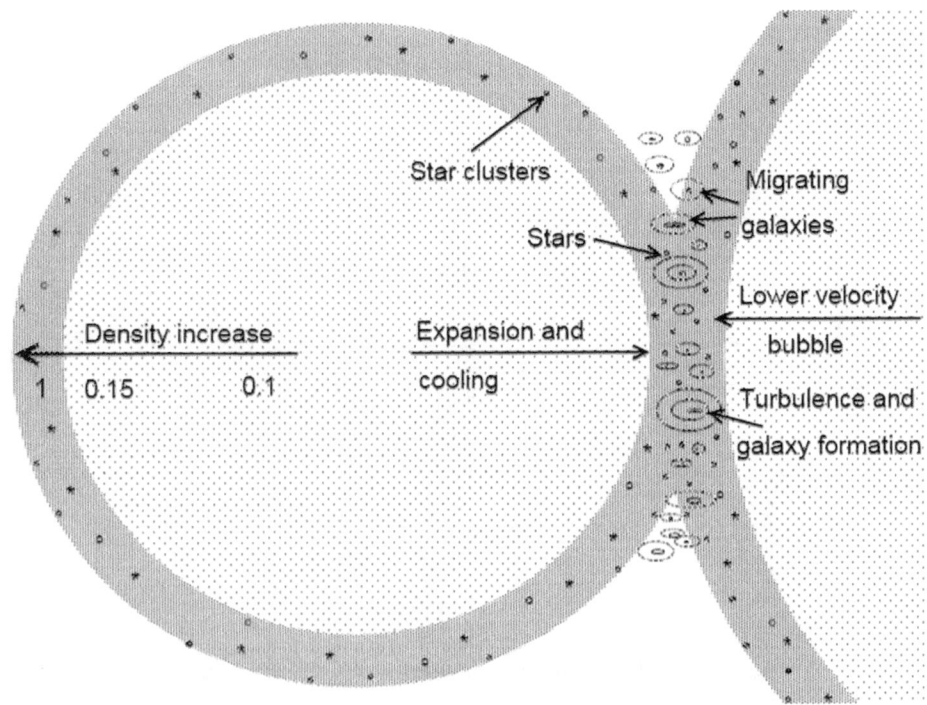

Figure 4.19 Turbulence and galaxy formation between two bubbles

As the two shells interpenetrate each other to form a common wall, their respective kinetic energy is very high, with velocities of a few thousand km/sec each, and this will produce high gas compression with large hurricane-like turbulence and eddies in a generally expanding and cooling environment so that the formation of galaxies will gradually proceed as the wall closes in with a flat or curved shape depending on the respective bubble size. Then the residual kinetic energy left after the curling and whirling of galactic formation will push galaxies at lower velocities (streaming motions) out of the wall toward the concave triangular three-bubble junction from which it will gradually be squeezed out toward a concave tetrahedron at the junction of four bubbles to form a small to average-size galactic cluster that will in turn be squeezed out and pulled out to its final destination in a concave hexahedron at the junction of six bubbles where a large cluster made up of galaxies and hot gas will combine an equivalent critical mass of about seven thousand galaxies will gather in an ultimate black hole like a quasar and start a new local cosmic cycle in a maxibang implosion-explosion.

Figure 4.20 Wall formation between two expanding bubbles

Galactic formation between two expanding bubbles is presented in figure 4.20, where a light ray is shown crossing the two expanding volumes and the central wall. Turbulence is present along the wall for the formation of galaxy groups like our own galaxy with Andromeda and surrounding galaxies. The thickness of the wall and the spread of galaxies away from the center of the wall will vary according to the curvature of the wall and the velocities of the clashing bubble shells.

But galaxy formation will follow a sequence starting with high energy eddies generated by high kinetic energy from the expanding gas that will produce the first stars and small galaxies that will be gradually incorporated in more massive alpha galaxies after being stripped of their gaseous material, some ending as globular star clusters and others as irregular dwarf galaxies. Arrows indicate redirected bubble expansion and the outflow of gas and galaxies toward the clusters. Note that the area near the wall may represent a sector where expansion is diverted sideways and more likely decreased, absent, or slightly inverted, thus inducing some errors in distance evaluations based on redshifts. For example, in relation to the cosmic microwave background reference frame, it is estimated that our Milky Way galaxy is moving at a velocity of about 600 km/s in the direction of the Hydra Centaurus supercluster located 150 M l-y away. This movement may be more related to the residual kinetic energy of bubble expansion rather than solely to gravity attraction by the mass of a large cluster.

4.2.2 Contraction Space or Blueshift between Three Bubbles

As we progress from simple to more complex models, many questions arise, like what happens when a light ray crosses three bubbles and two curved walls? After evaluating that bubbles include about 90% of cosmic volume, where half the bubbles are likely expanding and redshifted and the other half have ceased to grow and are not redshifted, this leaves about 10% of the space for clusters, walls, and filamentary structures. This means that about 10% of cosmic volume is probably not redshifted; on the contrary, this 10% is more likely neutral, or blueshifted. Therefore, a good cosmic accounting will include in its equations a systematic low redshift, neutral, or blueshift correction to obtain the equilibrium between expansion, neutrality, and contraction with adjusted shift z, corrected distances, and averaged Hubble rate of expansion.

Figure 4.21 Expansion and contraction between three bubbles

The expansion and contraction among three expanding bubbles are schematized in figure 4.21, where a larger central residual bubble is flanked by two smaller and similar bubbles [30]. A dot indicates the original center of each bubble that is crossed from right to left by a ray of light. Arrows are roughly proportional to the velocity vectors of expansion. As the two smaller bubbles collided with the larger one in the center, most of the kinetic energy was transformed to curls and twirls from which sprung stars and galaxies along the common walls. The smaller bubbles expand at a higher velocity and related kinetic energy so that the common wall is curving inside the larger bubble per equation 2.1. For example, original radial bubble velocity could be 2545 km/sec for the two small ones and 2290 km/sec for the larger one.

Each small bubble has a width of 40 Mpc and a redshift value $z=0.00933$ representing a positive expansion velocity $cz=2,800$ km/sec and a standard Hubble rate $H=70$ km/sec/Mpc. But the light crossing the central bubble corresponds to a distance of 55 Mpc and a blueshift value $z=-0.0017$, which represents a negative expansion velocity or contraction $cz=-510$ km/sec (2X-255km/sec) and a low negative Hubble rate $H=-9.3$ km/sec/Mpc.

Now we may compute the total for a light ray crossing these three bubbles: the total distance is 135Mpc with a net redshift value $z=0.01696$ representing a total expansion velocity $cz=5,100$ km/sec and a low Hubble rate $H=38$ km/sec/Mpc. But if an astrophysicist uses a different model, like the big bang, and takes the net observed redshift value $z=0.01696$ to compute a total expansion velocity $cz=5,100$ km/sec and divide by the average Hubble rate of expansion $H=70$ km/sec/Mpc, then he would obtain a distance of about 73 Mpc instead of a total of 135 Mpc, which means that only 54% or about half of the total distance would be recognized so that the other half may be invisible and neglected in such a model based on the concept of a universal Hubble constant from a unique big bang event instead of multiple bang events averaging to the Hubble constant.

Most of the large volume of the central bubble on the figure, between points numbered *2* and *3* along the light ray, would not be visible in redshift space mapping. Galaxies in wall *2* would appear slightly farther than those in wall *3*. Thus, galaxies in wall 2 and 3 would seem close to each other and have their relative distance inverted. A double wall with double groups nearby would be observed instead of the two common walls drawn in the figure. The consequence is that we should find low redshift, neutral, or blueshifting bubbles in narrow or null spaces between the more dynamic and expanding bubbles with a larger redshift. This explains the appearance of large double pseudo-walls in redshift space as seen in figure 4.1 and in all slice or 3-D surveys across the cosmos.

4.2.3 Tetrahedral Space between Four Bubbles

The tetrahedral space between four spheres and compacted bubbles is illustrated and compared in figure 4.22. The sketch on the left is similar to figure 2.7, but with the top sphere lifted to show the heart of the small cluster that would be nested in the center.

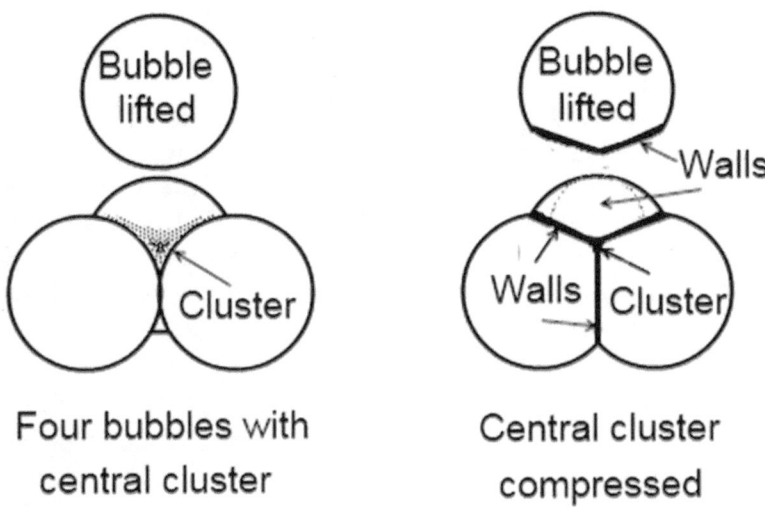

Figure 4.22 Tetrahedral space between four bubbles

On the right, the upper bubble is also lifted to show the reduced tetrahedral space in the center. The central cluster is compressed because the shells of the four bubbles have merged two by two for the formation of six walls and four concave triangular ducts that are guiding the spinning galaxies toward the more massive central cluster. The shape of these small tetrahedral clusters is illustrated in figure 4.23. In a real cosmic example, the shape of the tetrahedron will differ since the four bubbles are most likely of different size and not symmetric.

Figure 4.23 Cluster in acute tetrahedron

On the left of the figure, we present a view where all four apexes are well-defined and where the concave shape of the four sides is shown. A different view from the top of any apex is presented on the right side, with the concave curvature visible at the base. These tetrahedral structures generally correspond to smaller intermediate clusters that will eventually merge with the larger hexahedral clusters. They contribute to the knotty weblike structure of the cosmos by their aspect.

Figure 4.24 Active galactic nucleus (AGN) in contracting tetrahedron

A 2-D sketch of a contracting tetrahedron is presented in figure 4.24, with a large galaxy near the center harboring an active galactic nucleus, or AGN, for short. We expect that most tetrahedral galactic clusters should have such an active nucleus when a number of galactic groups have started to accumulate in the center from the residual kinetic energy and pressure of the expanding bubble. Some of the infalling galaxies will be caught in the gravity field of the more massive center and may either collide and merge or gravitate for a period of time around the larger mass.

The virial theorem states that if equilibrium is present in a gravitational attraction system, such as a cluster of galaxies, then twice the average kinetic energy equals the average potential energy. But in our model, equilibrium for a virial radius may only be partly reached for galaxies gravitating close to the central mass of the cluster since other incoming galaxies are endowed with residual kinetic energy from bubble expansion. Thus, the high velocities of infalling galaxies are only partly related to the cluster's gravitational center and useless to evaluate the total mass of a cluster, so dark matter is unjustified on this ground [57].

4.2.4 Hexahedral Space between Six Bubbles

We have seen in the Kepler conjecture that every sphere is surrounded by six hexahedral openings, and since each of the latter space is made up by the intersection of six spheres, we should find on average one hexahedron per sphere.

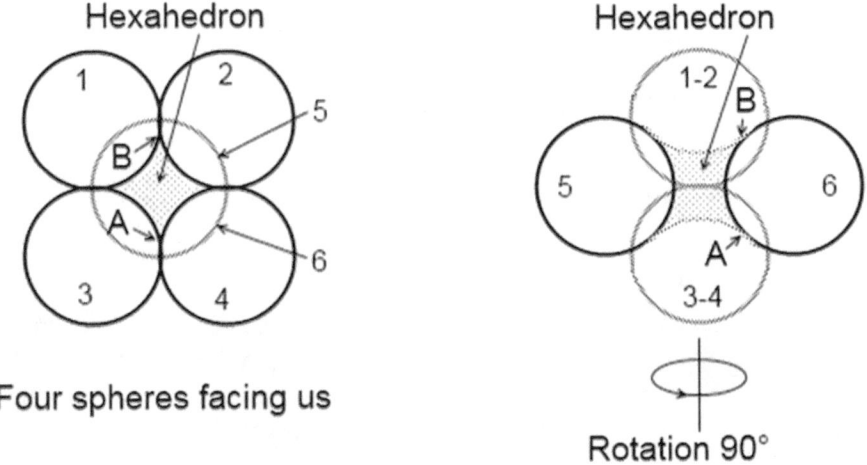

Figure 4.25 Hexahedron between six spheres

Figure 4.25 represents two different views from the hexahedron located between the spheres numbered 1 to 6. On the left, sphere number 5 is in the back, and number 6 is in front. The sketch on the right corresponds to a 90° counterclockwise rotation of the left drawing as seen from above: spheres 1 and 3 are respectively in front of spheres 2 and 4 after rotation. Letters *A* and *B* are used as markers to locate two different apexes before and after rotation.

Of course, the kinetic energy of bubble expansion will compress hexahedrons to a smaller size between six bubbles, while galaxies are forming and migrating into these high-density clusters. Since the topology of all large-scale structures in the cosmos always corresponds to imperfect or irregular geometric systems, we expect to find spheres and bubbles of various sizes, so pentahedrons, for example, may also be found in observations instead of the expected hexahedrons.

Figure 4.26 Concave hexahedron produced by six spheres

A blow-up of a concave hexahedron formed by six spheres is presented in figure 4.26. The concave curvature of the three sides facing us is shown. The shrinking from bubble expansion will vary according to the size of each bubble involved, and the hexahedron should eventually contract to a cluster where a quasar will form by merger of a large number of galaxies.

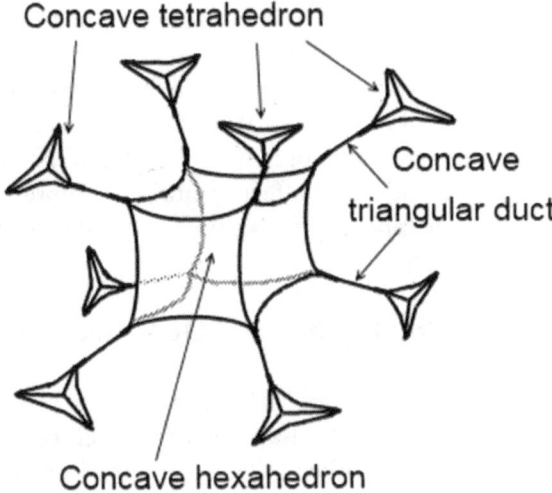

Figure 4.27 Concave hexahedron and feeders

In the face-centered close packing arrangement, a concave hexahedron may be fed by eight concave tetrahedrons via eight concave triangular ducts as shown in figure 4.27. We observe hierarchical structures leading to the hexahedron: groups of galaxies in walls between two bubbles followed by the triangular ducts and tetrahedrons. Note that two tetrahedrons located in opposite directions are replaced by two concave hexahedrons in the hexagonal close packing arrangement.

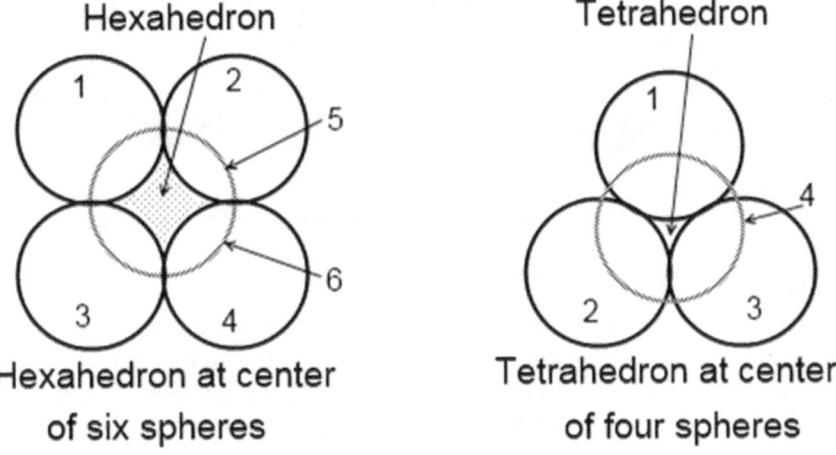

Figure 4.28 Comparison of hexahedron and tetrahedron size

A comparison at the same 2-D scale between hexahedron and tetrahedron is displayed in figure 4.28. We observe that the hexahedron occupies a volume many times larger than the tetrahedron. This explains why we expect tetrahedrons to form small migrating clusters and hexahedrons to contain the largest clusters with the development of large AGN and quasars that will eventually disappear under the effect of gravity compression as the most massive black holes and reappear as expanding new bubbles after a maxibang event.

4.2.5 Old Bubbles and Dodecahedron

But what happens to an old bubble that has expanded and clashed with as many other bubbles as possible? We have seen that each sphere in the Kepler conjecture is in contact with 12 other spheres, so we expect a mature bubble to have a maximum of 12 flat walls in common with its neighbors, thus forming a dodecahedron in the ideal geometric situation.

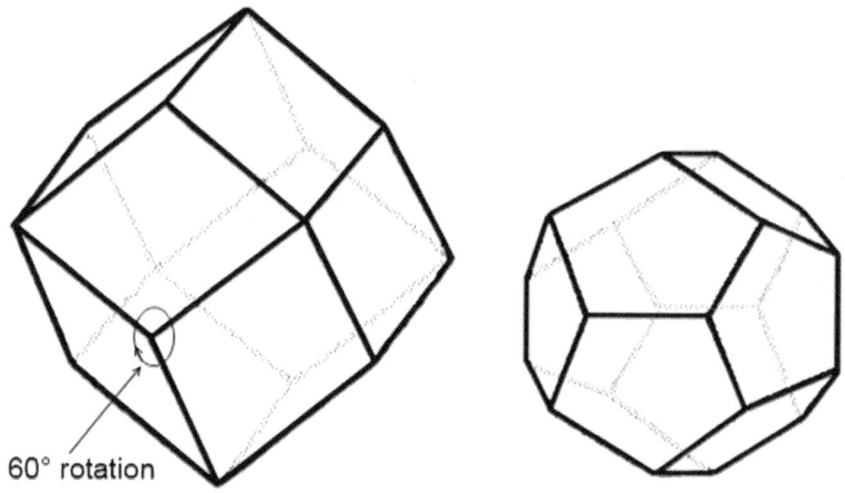

Figure 4.29 Dodecahedron or maximum number of bubble walls

Two different configurations of a dodecahedron are presented in figure 4.29. On the left, we find a dodecahedron composed of 12 squares and corresponding to the face-centered close packing arrangement of the Kepler conjecture. A 60° rotation of three of the square faces would yield the hexagonal close packing arrangement. On the right, the 12 pentagons forming a dodecahedron are not corresponding to the expected local topology of a mature bubble.

4.3 Expansion and Contraction Intervals

A sketch of light rays crossing many bubbles is presented in 2-D in figure 4.30. Smaller and younger bubbles grow faster than older ones with consequent higher redshift rates and a shorter-lived youth. On the contrary, older bubbles grow more and more slowly with lower redshift rates and an extended life when space is available. In the walls and junctions separating the bubbles where redshift growth has ceased, we can find blueshifting between merging proximal masses forming the galaxies and between the groups and clusters. After a bang, every new bubble takes up some space volume from up to 12 old bubbles. There is approximately one maxibang in the cosmos every 3.5 years, but the probability to see one is slim because of the slow speed of light over cosmic distances. So we could perhaps witness one in 3,000 years in the readily accessible part of the cosmos surrounding us if we keep track of all presently known quasars during that time.

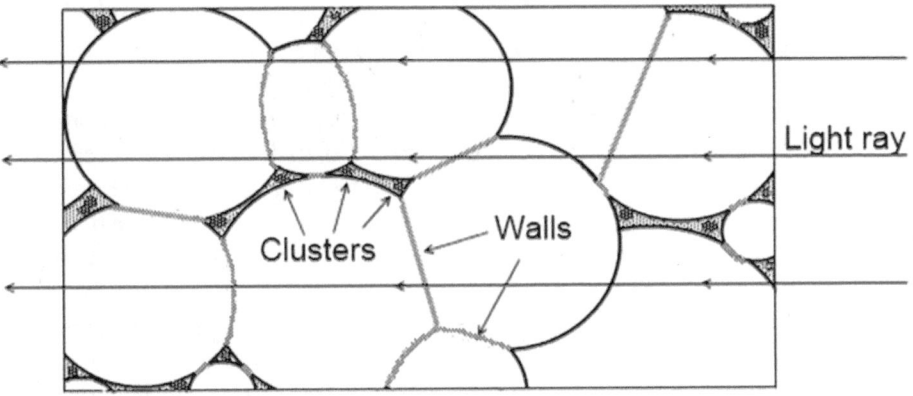

Figure 4.30 Sketch of light rays crossing many bubbles

As a permanent feature, small bubbles keep expanding into older bubbles with their high kinetic energy and clashing with younger ones to form new galaxies where starlight will start on its cosmic journey. Light coming from faraway galaxies will generally cross groups of two or three bubbles in close contact because of a tendency for the repetitive presence of lattice layering of two or three layers across the cosmos, as previously shown in the Kepler conjecture for face-centered and hexagonal close packing arrangements.

4.4 Distance and Redshift

What could be a realistic constant for distance correction q between big bang and multibang models? The scale of the distance shown in the 2dFGRS data slice of figure 4.1 was evaluated at 0.92 billion l-y for a redshift z=0.1. But the more recent evaluation of distances in the SDSS indicates a value of 1.3 billion l-y for a redshift z=0.1, which implies an increase of about 41% over the 2dFGRS estimate. This is not negligible and seems a good indication that distances are underevaluated in the big bang model. We saw in our example in figure 4.21 that, in a three-bubble model, an error in interpretation could yield a distance nearly half the real value in the model. In both cases, the range of correction or error is large and varies between 40% and 85%. So a first tentative evaluation of constant q for very large cosmological distances could be in the range $1.40 < q < 1.85$ [54].

Note that when doubling distances with q=2, volumes and masses are increased by a factor of 8, the cube of the radial distance. Thus, other methods of distance evaluation, such as the empirical relationships from Tully-Fisher for spiral galaxies or Faber-Jackson for elliptical galaxies, should be combined to the standard candles as checkpoints in the distance ladder.

The satisfactory measure of distances is one of the most basic challenges in astronomy. Many methods are combined to build the cosmic distance ladder. Each method is used as a rung in the ladder. The closer one and most precise is based on geometric parallaxes measured from the extremity of the earth orbit or from extreme satellite orbits. It yields distances of up to 100,000 stars in our galaxy and distances of up to 3,000 l-y.

An approximate distance may be computed using the apparent brightness of a star, then assume its luminosity and apply the inverse square law. A distance of about 300,000 l-y may be reached if we study a star spectrum to obtain the temperature and then extract the luminosity distance from a Hertzsprung-Russell diagram.

The next step up the ladder is based on the fixed or variable but predictable luminosity of some objects, such as the RR Lyrae and Cepheid variable stars and the supernovae Ia (SN Ia) explosive stars. A distance of three million l-y may be reached with the RR Lyrae and of about 100 million l-y with the Cepheids. But with the SN Ia stars, a much larger distance of 30 billion l-y may be reached as a luminosity distance estimate if we neglect selection effects and secondary effects with large distances, such as reddening, absorption, etc.

4.4.1 Redshift and the Moving Sidewalk

The following example in figure 4.31 should help to understand how the effects of space expansion, such as redshift and stretched wavelength, may be produced in a nonexpanding system that mimics expansion.

Figure 4.31 Jim walking in the opposite direction to the moving sidewalk

Imagine a 0.9 km long moving sidewalk like those found in airports and Jim always walking at a constant speed of 3.6 km/hr or 1 m/s both on and off the conveyor belt. His constant speed of 1 m/s before reaching the conveyor represents steps of 1 m, each corresponding to the wavelength, and one step or cycle per second representing the frequency. If the belt was not moving, Jim would cross its 900 m length in 900 seconds or ¼ of an hour. But the walkway is moving in the opposite direction at 1.8 km/hr or 0.5 m/s, so he needs twice the time, or ½ hour, instead of ¼ hour to cross 0.9 km. Thus, in ½ hour, he moves forward 1.8 km and backward 0.9 km for a total distance of 2.7 km, or 2,700 m, which is divided by ½ hour, or 1,800 seconds, and yields a stretched wavelength or step of 1.5 m. When Jim's speed of 1 m/s is divided by the observed wavelength equivalent of 1.5 m, we find a frequency of 0.66 steps or cycle per second. Then if we divide the belt speed by Jim's speed, we find the equivalent of a redshift where z=0.5. The moving belt has the same redshift effect as expansion although nothing expands.

4.4.2 Cellular Expansion Model and Redshift

Stable growth and renewal of cells in nature may be compared with our cosmic self-regulatory model in which older bubbles are replaced by newly born and expanding bubbles whenever gravity takes over and crushes a large galactic cluster to rebirth.

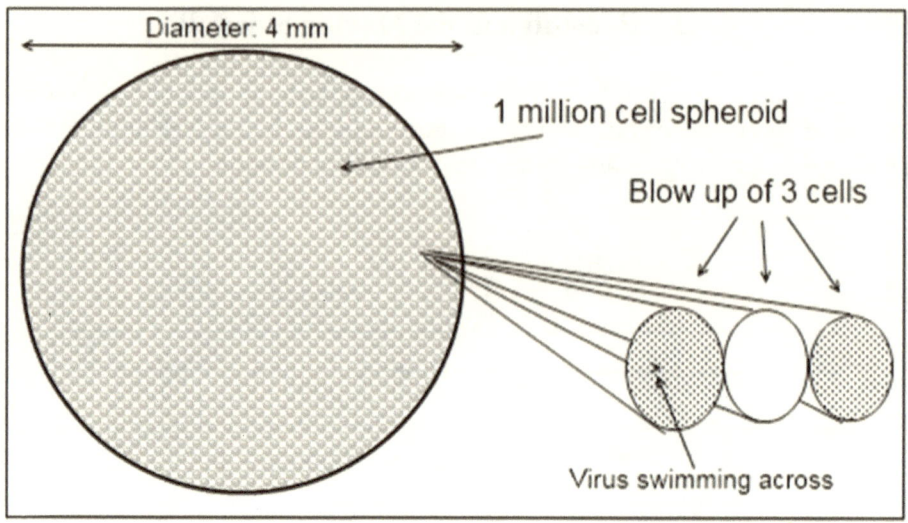

Figure 4.32 Self-regulation of growth cells in a 3-D spheroid

Laboratory studies on the growth of tumor cells in 3-D has demonstrated that a maximum diameter of about 4 mm and the formation of about one million cells may be reached for isolated spheroids [33]. Then self-regulation takes over by controlling nutrient supplies to the far center, but new cells continue to replace older dead cells within the spheroid.

Now imagine that half the cells in the spheroid are growing and that the other half is being recycled in such a way that equilibrium is maintained both in spheroid size and in the total number of cells, then a virus swimming at constant speed across the spheroid would take a longer time to cross all the growing or expanding cells, one out of two, and a normal time to cross the other half that is not expanding. If we assume that material is recycled along the walls between cells, then very little blueshift equivalent would be associated with the wall crossing.

Therefore, the expanding half of the cells would produce a redshift equivalent or a sum of small redshifts and would only give the measure for about half the distance across the spheroid. This suggests doubling distance measurements, so a larger value for constant q, closer to q=2, is suggested. Note that our concept of bubble has some similitude to cells growing and dying in a spheroid with common walls and recycling. Of course, the cosmic concept has little to do with the exploding or popping action of soap or other gaseous bubbles. But we can conceive a cosmos in equilibrium where about half the space is filled with bubbles surrounded by expanding shells, while the other half is made of nonexpanding dead or dying bubbles whose shell's material is squeezed out along the walls into the clusters.

4.4.3 Lookback Time, Distance, and Redshift

Photons are the fastest calibrated wave particles traveling at the speed of light c in the cosmic vacuum. Their speed c is multiplied by the number of years to obtain galactic and intergalactic distances when no stretching is present. Since 1929, when Hubble discovered a rough correlation between distance and galactic recession velocities, many observations and writings have covered the subject with sometimes large divergences. Presently, most big bang models include a Hubble constant, dark matter, and dark energy. Our multibang model is simpler and is based on observations: no dark matter and no dark energy since the latter two are unjustified and have never been directly observed or proven.

Figure 4.33 Lookback time or distance in relation to redshift z

Over very large distances of the order of z>1, we find in the multibang model a direct correlation between redshift z and distance or time in the past as illustrated in figure 4.33. But the big bang model only accounts for a maximum cosmic age of 13.8 billion years and a basic distance of 13.8 billion l-y, plus a cosmic distance expansion based on a stretch factor proportional to the Hubble constant H_o, for a total representing only a fraction of the distance to faraway objects, thus skewing the results. For example, the big bang distances are smaller by a factor of 1.31 at z=1, and by a factor of 8.0 at z=20, the latter redshift corresponds to the expected cosmic diameter in the multibang model.

Such a discrepancy between the two models has consequences on the interpretation of quasar spectra, dust distribution, and all high z signals from galaxies and clusters. For example, in the multibang model, we expect that a photon coming from a z=1 emitter will not only be redshifted but cross about 110 bubbles corresponding to 110 bubble walls with the associated risks of scattering, absorption, and reddening from gas clouds and dust and lensing from massive galaxies and clusters. In the big bang model, a maximum of only 84 bubbles would be expected for the same redshift.

But for a redshift z=20, the multibang model estimates that 2,200 bubbles with bubble walls would be crossed as compared to a maximum of only 275 bubbles in the big bang model. This is a major difference in the big bang model, 4.5 times less bubble walls at z=10, 8 times less at z=20, and so on, which indicates an undervaluation for gas clouds and dust across the cosmos with an increasingly skewed model with distance, implying gradually lower distances than the multibang as shown in figure 4.33.

4.4.4 Luminosity Distance and Redshift

In a multibang universe where bubble expansion is present, distant light sources are expected to be slightly dimmer than would normally be found with the simple inverse-square law because of all those line-of-sight interferences, such as diffractions, lensing, absorptions, etc. This is exactly what we will discover on figure 4.34 using supernovae Ia luminosity-distance: that these supernovae are slightly dimmer and clearly true and reliable standard candles in the cosmic distance ladder since they fit with the multibang time and distance curve with no expansion and no acceleration. But in a nonpermanent big bang universe that is born in time and expands, the distance and time measured with the speed of light as a ruler is limited by the present cosmic age and the exponential contraction of events backward in a finite space-time, so distant light sources are expected to be brighter, which is not the case as observed from supernovae and other sources.

Figure 4.34 Low-z cosmic time and distance

A graph of low-z cosmic time and distance is presented in figure 4.34. The horizontal scale shows the redshift between z=0.1 and 2.0, and the vertical scale represents time or distance in Giga light-years. This graph is a blowup of the two first values of figure 4.33 to show the low-z divergence between the three lower curves when events are located closer to the observer. The big bang luminosity distance up to z=1.3, calculated from supernovae Ia, is presented as a dotted curve. The multibang luminosity-distance curve, also calculated from the same supernovae Ia data but after removal of the big bang expansion stretching, is represented as a dash-dot-dot curve. The solid black line is the multibang time and distance curve. The big bang paradigm for the two lower curves is based on a flat model with a Hubble constant Ho=71 km/s/Mpc, the density parameter for baryonic plus dark matter is Ω_m=0.26 and for the dark energy is Ω_Λ=0.74, where distance is expressed as the dash curve and time as the dash-dot curve. The two multibang values being the same, four values are indicated at z=0.1 for the curves in the respective order.

Since the standard candles used to measure large-scale cosmic distances are based on supernovae light curves and extend approximately to z=1.3, we can see that the divergence between the two evaluations become increasingly important with higher redshifts and may skew distance measurements based on the big bang model. But what big bang model, since the curves are bending in two different directions on both sides of the multibang model?

Two projects with redshifts z<0.9, the Supernova Cosmology Project and the High-z Supernova Search Team, announced in 1998 the unexpected discovery that their study of high-z type Ia supernovae (SNe Ia) found them dimmer than expected, which implied that they were farther away in the big bang model. Moreover, their results show, with very large error bars, a correlation between the widths of the light curves and (1+z), which they interpreted as a strong argument in favor of relativistic time dilation related to the expanding universe.

The two teams erroneously interpreted the dimming as a sign that the expansion of the universe was not steady or slowing as expected but accelerating and that the acceleration was caused by the presence of ad hoc invented dark energy filling up space and accounting for more than two-thirds of the density in the universe. They related this negative pressure from the hypothetical dark energy to the cosmological constant lambda that Einstein had rejected on the count of cosmic expansion after introducing it in his equations as a mean to counteract gravitational compaction and explain his balanced cosmic model.

Since 1998, a few additional surveys of higher redshift SNe Ia have been interpreted as indicating not only an acceleration but also a deceleration for higher z values in order to explain the stronger attraction in the more compact big bang model that would prevail in the early expanding universe. A review of 306 quality SNe Ia data up to a redshift z=1.6 was published (Kowalski et al., 2008) with similar conclusions.

The same quality data was reinterpreted as consistent with a static universe (Crawford, 2009) [88] using the argument that the increasing widths of the light curves in proportion to (1+z) are related to selection effects since previous studies used the peak magnitude as the standard candle instead of the more reliable total energy of the supernovae explosions. This paper represents a strong support for a model with no time dilation and consequently no universal expansion. Also, no time dilation was found in a study of gamma ray bursts (Crawford, 2009a, 2009b). Another paper (Hawkins, 2010) [89] analyzed the Fourier power spectrum from the light curves of about 800 quasars equally spread on both sides of z=1 and found no time dilation effects, which contradicts the big bang model.

Time dilation is a universal phenomenon associated with the relative movement of any object. It is proportionately more important in Lorentz-Einstein equations when the relativistic velocity of a particle is close to the speed of light. The clocks of earth satellites are slower and constantly readjusted, and large corrections are found for the slower clocks of particles in man-made accelerators.

In the multibang model, time dilation is certainly present and generally negligible across any expanding bubble and will depend on the relative position of each object between different bubbles. No acceleration is present other than the local intra or interbubble relationships, and thus, no overall cosmic acceleration or dark energy is justified. The larger multibang time and distance curve in figure 4.34 corresponds to more realistic and calibrated values for dimmer SNe Ia than those from the big bang curves. The luminosity distance with slightly larger distances than the multibang model, may be associated with the gradually higher amount of reddening, absorption, etc. And since it is easier to see the brighter distant supernovae, it may also stem from small selection errors.

In conclusion, the dimming of SNe Ia fits the time and distance curve from the multibang model without cosmic expansion and acceleration. Therefore, SNe Ia supernovae are standard time and distance candles with standard luminosity as expected in the latter model.

4.5 Cosmic Microwave Background (CMB)

The cosmic microwave background or CMB [37] is the cemetery of all emitted photons from all sources that have generally traveled many times back and forth across the cosmos and have stretched by an average z factor of about 1,100 over billions of years. The CMB is the blackbody signature of all EM energy from the cosmos that may be identified as the ultimate black hole from which nothing can escape, not even gravity. At the peak, it is the state of equilibrium between energy input from inverse Compton scattering and output from redshifting reached by photons in transit between the far infrared and the radio frequencies from which no additional heat is extracted.

Among others, cosmic rays, high energy particles from active galactic nuclei or quasars, and the Sunyaev-Zel'dovich effect play an important role in feeding back some energy to the CMB and preserve equilibrium [18]. It is the most important background radiation with a density of about 400 million photons per cubic meter [55]. The CMB is the most perfect blackbody since it includes all the electronuclear energy of the cosmos that is generated within the cosmic multibang spheroid and recycled in black holes by gravity compaction.

The WMAP microwave anisotropy survey from NASA has mapped the CMB to the best coverage and precision up to the year 2010. Power spectrum maps showing a larger peak and smaller repeats of temperature fluctuations at the micro Kelvin scale and based on the angular scale of anisotropies have been published and interpreted in various ways by the tenants of the big bang model. They generally interpret the large peak near an angle of 0.9° as representing all the cosmic mass energy, the second peak as all the baryonic matter, and the third peak as the ordinary plus dark matter.

In the multibang model only baryonic matter is present, and the large peak at an angle of 0.9° is associated with the presence of the large clusters of galaxies with hot gas where the Sunyaev-Zel'dovich effect prevails; angles larger than 0.9° are related to the variable size and distribution of bubbles as delimited by walls and clusters and thus are related to the larger structures within the limits of resolution of the probe [82], and the secondary anomalies are related to secondary structures like smaller clusters and groups of galaxies, but since clusters are fed by triangular ducts away from the walls and thus are generally not spherical, the secondary anomalies may also be related to these smaller structural appendages.

The CMB is definitely not the result of a unique big bang event followed by photon dissociation from captured electrons that would fill the present-day sky. This is physically impossible since such an event occurring about 370,000 years after the bang would form an expanding shell radiating away all that energy to a time distance of about 13.7 billion l-y (G l-y), and since light travels at 300,000 km/sec, then the Hubble expansion rate at 71 km/sec/Mpc (1 Mpc=3.26 M l-y) would never be able to catch up with the faster-expanding shell of the runaway light, and with an open or flat model, there is no mechanism to return the light into the expanding cosmos. Also, there is no scientific or satisfactory basis for an inflation scenario except to justify a particular model. But luckily, there is an alternative.

4.5.1 Black Body Spectra for Three Temperatures

Three different blackbody spectra are drawn in figure 4.35. The first is for a temperature of 25,000 K representing a hot blue star at the top, followed in the center by 3,000 K corresponding to the most common orange-red dwarf stars, and by 2.725 K from the cosmic microwave background.

Figure 4.35 Blackbody spectra for three temperatures

Our sun, with a surface temperature of about 5,800 K, would fit between the 3,000 and 25,000 K curves with a maximum in the visible range. A wavelength logarithmic scale is displayed at the bottom, and a frequency scale in Hertz is found on the top. The narrow gray band in the center corresponds to the visible part of the spectrum. It is flanked on the left by ultraviolet and shorter wavelengths and by infrared to radio wavelengths on the right.

The height of the curves is proportional to brightness temperature usually expressed as power per unit area, for example, in watt per square meter (W/m²), which may also represent a density of energy or intensity function in the Planck radiation law applied to a blackbody spectrum. The CMB maximum for Planck's law is at a wavelength of 1.873 mm, corresponding to a frequency of 160.2 GHz. But the Wien displacement law states that the maximum wavelength in micrometers is equal to 2,897 divided by the temperature in K, which is $\lambda_{max} = 2,897/T_K$. So the CMB maximum shown on the Wien's slope as a gray line is at a wavelength of 1.06 mm, which is a frequency of 283 GHz. We observe this displacement of the maximum blackbody emission from the three spectra in figure 4.35 that varies inversely with the blackbody's temperature.

Thus, the relation between Planck's and Wien's maxima is not linear, and the maximum from a colder object corresponds to a longer wavelength.

4.5.2 Extragalactic EM Waves Background

The blackbody curve from the Cosmic Background Explorer (COBE) and Wilkinson Microwave Anisotropy Probe (WMAP) satellites with the millimeter peak of the CMB is the same curve as that of the CMB blackbody spectra in figure 4.35 but at a different scale and is reproduced in figure 4.36, with the wavelength and frequency scales at the bottom and the energy scale in electron volt (eV) at the top [61]. The y-axis represents relative brightness with a log scale.

This sketch is a summary of the more comprehensive distribution of all extragalactic EM waves that have been so far recorded by a variety of instruments sensitive to different wavelengths or photon energy. Visible light corresponds to the narrow vertical band. Note that the relative brightness of the cosmic ray particles (not shown) may reach a level similar to the CMB but in the energy range of the gamma rays. The most energetic cosmic ray particles should be found to the left of the gamma rays with a gradually lower relative brightness.

Figure 4.36 Extragalactic EM Waves Background

The continuous black band represents a rough estimate of the cosmic background spectrum of all EM waves, including the CMB spectrum. The cosmic gamma ray background (CGB) is relatively continuous and represents mostly energy from active blazars and other high-energy sources. The big bumps in the cosmic X-ray background (CXB) is generated by all active galactic nuclei and by the clouds of hot gas in galactic clusters. The cosmic ultraviolet (CUB) and optical background (COB) are produced by the EM output from the higher frequencies of all the stars and galaxies, and are more strongly attenuated by absorption at the high ultraviolet frequency end. A gap between optical and infrared needs further and more sensitive studies. The cosmic infrared background (CIB) generated by orange-red stars, heated dust, and gas clouds is usually associated with the small far infrared bump at the foot of the CMB. The peak of the CMB is about 30 times brighter than the far infrared. Note that the near infrared is included with the optical background. Finally, the cosmic radio background (CRB) corresponds to a low brightness, but due to absorption, little is known about the longer radio wavelengths.

The common problem associated with all these background studies is the difficulty to remove or subtract light pollution from zodiacal sources in our solar neighborhood and from proximal galactic sources to obtain a clean background. A minimum threshold in signal-to-noise ratio must be reached before confirming the brightness of a specific background frequency.

This smoothed EM spectrum of the thermal and nonthermal radiation in the cosmos shows, except for the radio background, a general rise in the curve from high-energy to low-energy photons. Since the high-energy gamma ray, X-ray, and ultraviolet photons are more reactive, they travel shorter distances and are mostly absorbed in gas clouds or space dust and are reemitted at lower frequencies in the optical and infrared, thus increasing the relative brightness of the background toward longer wavelengths.

The infrared wavelength photons may be stretched for hundreds of billions of years, as they generally have a longer mean-free path across the cosmos, until they reach the millimeter microwave background where they accumulate to a higher density from the Sunyaev-Zel'dovich effect and until equilibrium is reached with the amount destroyed and recycled by high energy particles and by black holes sipping them up. The accumulation at the CMB peak is like a traffic jam produced by the repeated frequency blueshifting of the microwave background photons over billions of years by inverse-Compton scattering, more specifically by the Sunyaev-Zel'dovich effect in which about 1% of the microwave photons are repeatedly energy boosted while crossing every hot electron cloud in galactic clusters.

4.5.3 Inverse Compton Scattering and the Sunyaev-Zel'dovich Effect

The inverse-Compton scattering is illustrated in figure 4.37, where photons from the microwave background are blueshifted when they get boosted by high-energy electrons in clusters of galaxies containing hot gas and this in proportion with the density and thickness of the gas in the cluster [23].

Figure 4.37 The Sunyaev-Zel'dovich effect across a cluster

The Sunyaev-Zel'dovich effect is a consequence of this shifting to higher frequencies of the microwave background photons, whereby a deficit in the number of photons is recorded for frequencies less than 218 GHz (wavelengths from about 15 to 1.38 mm) and an excess is shown for frequencies above 218 GHz (wavelengths from about 1.38 to 0.3 mm). So this small distortion on contour anomalies in micro K will make the CMB appear gradually fainter toward the center of a cluster at frequencies below 218 GHz and gradually brighter toward the center at frequencies above 218 GHZ.

This effect is independent of redshift since it shows with equal intensity in similar-size clusters regardless of the z value of the clusters, but it is limited by instrument resolution for high z values. Consequently, the Planck satellite is searching for galaxy clusters related to this effect, up to a limited redshift of z=1, with nine channels of frequencies both above and below the crossover frequency of 218 GHz.

A paper based on these Planck satellite data (Huriet et al., 2013) [97] shows clearly that the observed crossover frequency is constant and that no increase of temperature is properly justified in relation to increasing z, and consequently, the temperature of the cosmic microwave background, T_{CMB}, is constant everywhere in the cosmos at 2.725 K as proposed in the multibang model in which energy is conserved while redshifting per equation 3.8. But the big bang proponents want to save their hotter-in-the-past cosmic expansion model and have decided to scale this constant temperature with a factor (1+z) in order to show a temperature increase with distance in proportion with the redshift, arguing that redshifting decreases the background temperature by the same factor (1+z) at the cost of denying conservation of energy.

4.5.4 Trace Stacking and Redshift to a Blackbody Spectrum

In figure 4.38, we present five signal traces and a stack, which is the sum of the traces divided by the total number. The process of stacking gives a surprisingly smooth result because data of opposite signs cancel out and noisy data spikes are attenuated by the number of signal traces.

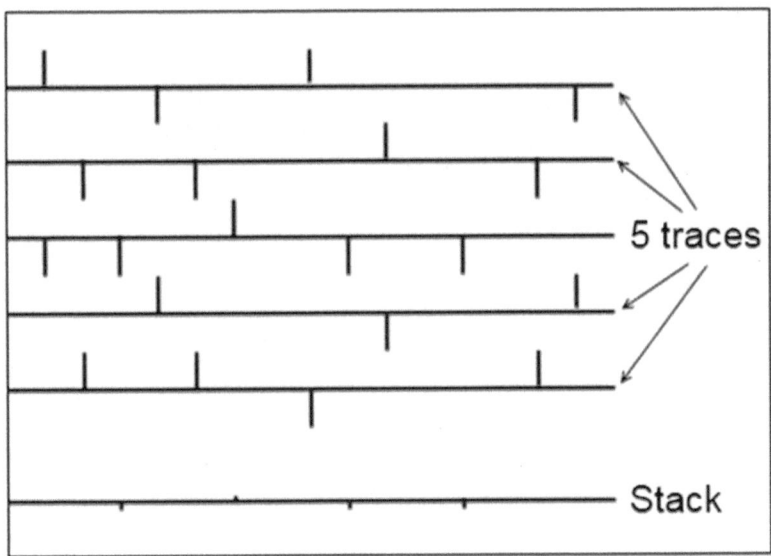

Figure 4.38 Stack of five signal traces

In nature, the blackbody spectrum of every star emission corresponds to a data trace with some absorption troughs that will add up to the signal from all the stars and emission peaks from gas clouds to yield after a long life of redshift the smooth cosmic microwave background where temperature variations are about one part in 100,000. These small variations are repeatedly maintained by inverse-Compton scattering back and forth across the cosmos. The repeated scatterings boost the brightness of the CMB to a value estimated to be about 30 times higher than that of the far infrared background.

A small part of the CMB escapes by redshifting to lower EM frequencies in the radio range, and equilibrium is also maintained by the continuous black hole sipping of CMB photons. In fact, the CMB represents the stack of redshifted blackbodies equivalent to the output of one trillion times one trillion orange-red stars in the cosmos. It is glaringly easy to see the outcome: an extremely smooth redshifted blackbody background that is continuously boosted by inverse Compton scattering up to its maximum and then degraded to radio wavelengths by a photon deficit.

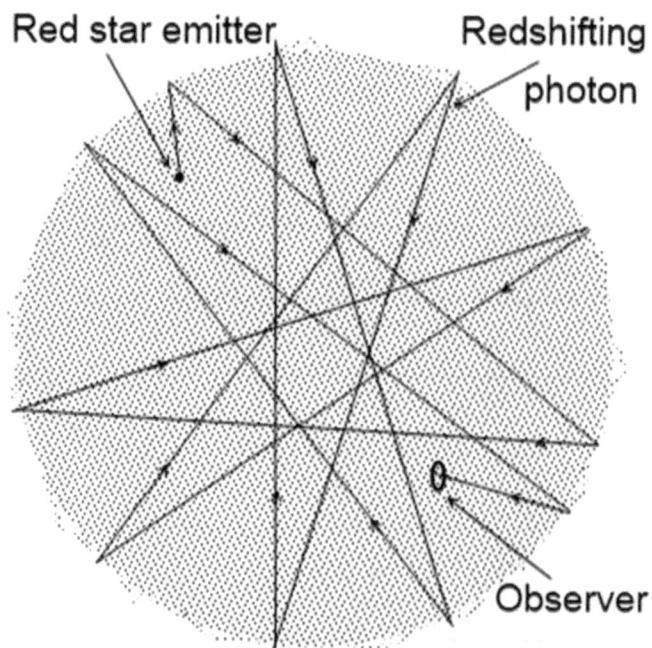

Figure 4.39 Trip of photons redshifted to CMB across the cosmos

A sketch of the trip of a redshifted photon back and forth across the cosmos until the CMB is reached is presented in figure 4.39. A typical red-star emitter is shown with a part of the trajectory to the observer. Of course, it is the travel across expanding bubbles that produces the redshift or wavelength stretch.

In fact, in the absence of any other interference, a 3,000 K average blackbody emitter will have its EM waves redshifted by a factor z=1,100 to the CMB at 2.725 K. In order to accomplish such a feat, and since the average cosmic radius of 143 billion l-y corresponds to a redshift of about z=10, then our photon must cross the cosmic diameter for an equivalent of about 55 times, which represents a total time distance of 15.73 trillion light-years.

In the absence of inverse Compton scattering, our photon would be redshifted directly across the CMB, from a wavelength of 0.3 to 15 mm, by a factor z=49, which represents 700 billion l-y, and the radio background would be of a higher relative brightness than the CMB since it would incorporate all the CMB energy plus all radio generating sources in the cosmos. Obviously, this is not the case since the radio background is way down from the CMB.

The risk of crossing a hot gas cluster is not negligible over such a distance and may occur many times to boost back the microwave background frequencies and maintain the central one at 218 GHz, which is about the average between Planck's and Wien's CMB maxima and also the transition frequency in the Sunyaev-Zel'dovich effect. Since the microwave background is a near-perfect blackbody, it means that all frequencies across this 2.725 K blackbody are equally blueshifted by inverse-Compton scattering over very large distances.

If we use the cosmic radius of 143 billion l-y, a bubble radius of 91 million l-y with two clusters per bubble, and the equivalent of seven billion clusters, each with a radius of 13.4 million l-y and each harboring an average of 500 galaxies, then a photon will hit only one cluster in 23 bubbles. We have, as in figure 4.13, but in 3-D, the relationship $2r^3=R^3$ where r/R=0.7937 within a sphere for the average distance crossing. This yields 1,764 bubbles that a photon will cross on average within the cosmic sphere or 2,200 bubbles across the full diameter. So each photon should react about 100 times across the cosmos for a redshift value z=20, or the equivalent of five hits for z=1.

But only about 1% of the photons react when crossing the hot gas of a cluster. Thus, each photon will react only once during a full trip across the cosmic diameter to maintain CMB equilibrium by adding photons between the limit of the infrared and up to 218 GHz and then removing photons all the way down to the onset of radio wavelengths. There are also limited secondary reaction contributions from other X-ray sources such as the high-energy AGN and quasars.

Figure 4.40 Comparison of 1 mm and 0.001 mm wavelength

What does a redshift around z=1,000 look like in terms of wavelength stretching and frequency decrease? The answer is illustrated in figure 4.40, with a comparison of 0.001 mm and 1 mm wavelengths shown at a scale 100 times larger. On the top, a frequency of 300 THz corresponding to the 0.001 mm wavelength is drafted as a very narrow vertical line measuring 0.1 mm in width. At the bottom, a frequency of 300 GHz corresponds to a wavelength of 1 mm that is drawn as a 100 mm wave. The impressive wave stretching is proportional to its factor of 1,000 in frequency decrease.

If we apply equation 3.8, E=h(1+z)f, where h=4.1357X10^{-15} e-V-s, to the nonstretched 300 THz with z=0 and to the stretched 300 GHz with z=1,000 frequencies, we obtain in both cases E=1.24 eV, which implies conservation of energy when bubble space expansion stretches EM waves across large distances within the cosmic spheroid.

5. Stars, Galaxies, Groups, Clusters, and Clouds

Stars are the batteries generating the electromagnetic and nuclear energy in the cosmos, while gravity is the power unit recharging them [22]. Galaxies are sweepers and concentrators of gas clouds that will generate the stars and thereafter crowd them up toward a central mass. The role of clusters consists mainly in building a nest of galaxies that gravity will eventually compress into an ultimate black hole and a new bubble cycle.

Even in the most imaginable static situation, the dynamics are overwhelmingly present from all the forces and associated fields acting and counteracting simultaneously with EM waves—photons, neutrinos, and all the mass energy present. So a static mass energy state should be seen more like a snapshot of a dynamic process or the temporary dynamic solid state of a fluid or plasma.

Fluid dynamics is that part of mechanics that concerns the motion of liquids, gases, and viscous materials including solids [45]. The behavior of a fluid is described by the Navier-Stokes equations, which are based on conservation of mass energy and of momentum, both angular and linear. A Reynold's number may be computed and assigned to a fluid to quantify the type of fluid flow: a laminar flow corresponds to a low number, and a turbulent chaotic flow with eddies will correspond to a large number.

5.1 Classification of Stars

Stars are divided in spectral classes designed by letters *O, B, A, F, G, K, M, L* and *T* in order of decreasing temperature, followed by numbers for each class from 0 to 9, which are increasing as the temperature decreases and described according to specific spectral characteristics that can be observed and measured from our earthly reference frame [77]. Table 5.1 is a summary of the spectral classes with a range of temperature, mass, radius, and lifetime. At both ends, an extra class is added for marginal or extraordinary stars, like class *W*, which represents spectral type Wolf-Rayet and in which stars are part of the hot giant class *O* stars but specifically hydrogen depleted and so in the process of burning helium and metals (metals in astrophysics are atoms larger than helium). At the other end, we find class *Y* in the process of being defined by astrophysicists near the questionable limit between stars and planets that is found at about 1% of the mass of our sun or 10 Jupiter masses and a star surface temperature that may be as cold as the human body, around 300 K, and thus definitely not a fusion combustion star.

Spectral class	Temperature in °K	Mass in sun masses	Radius in sun radiuses	Lifetime in million years	Relative fraction %	Mass X fraction%	Mass X percent X lifetime
W	50,000	50	20	0.5	0.0000001	0.000005	0.0000025
O5	40,000	32	18	1	0.00002	0.00064	0.00064
B0	28,000	16	7.4	10	0.03	0.48	4.8
B5	15,000	6.5	3.8	100	0.06	0.39	39
A0	10,000	3.2	2.5	500	0.3	0.96	480
A5	8,700	2.1	1.7	1,000	0.6	1.26	1,260
F0	7,500	1.4	1.4	2,000	1	1.4	2,800
F5	6,700	1.25	1.2	4,000	2	2.5	10,000
G0	6,000	1.04	1.1	10,000	3	3.12	31,200
G5	5,400	0.92	0.9	15,000	4	3.68	55,200
K0	4,800	0.8	0.8	20,000	6	4.8	96,000
K5	4,200	0.65	0.7	30,000	7	4.55	136,500
M0	3,500	0.45	0.6	60,000	10	4.5	270,000
M5	2,800	0.2	0.3	90,000	58	11.6	1,044,000
L0	2,250	0.09	0.2	92,000	5	0.45	41,400
L5	1,700	0.075	0.15	90,000	2	0.15	13,500
T0	1,300	0.04	0.13	85,000	0.91	0.0364	3,094
T5	900	0.025	0.11	80,000	0.1	0.0025	200
Y	400	0.01	0.1	75,000	0.001	0.00001	0.75

Table 5.1 Stars' spectral classes in terms of various parameters

In fact, the limit of hydrogen fusion is found near the middle of spectral class L, where a mass around 7.5% of the sun's is recorded and where fusion may be short-lived, only a few million years since convection cooling and mass losses may hamper any further fusion. But with only a slightly larger mass above this limit, L or M stars just above 8% of the sun's mass could theoretically live for one trillion years or more from a slow core fusion and whole body convection process, although in the table, we limit the expected lifetime to around 90 billion years since it is unlikely that these stars would survive more than two or three cycles from the local gravity crunches at intervals of about 25 to 30 billion years without being recycled. And no L stars are known to harbor fusion with a mass less than 7% of the sun's.

Surface temperatures vary from more than 50,000 K for the O stars all the way down to less than 400 K for the new class Y stars. Some of the temperature limits, and other characteristics vary in the literature as new data is coming in and interpretations are adjusted. Since the visible stars in the range of EM waves from 380 to 740 nm are the easiest to observe, they cover most of the table and subdivisions, but we should remember that the most numerous stars, at least in our galactic neighborhood, are from the fainter red dwarf M spectral class and have only 50% of the surface temperature, 33% of the mass, and 40% of the radius of the sun.

Stars' lifetime should vary from less than half a million years for the larger and hotter ones to possibly over 90 billion years for the slow hydrogen-burning red dwarfs, but the no-fusion brown dwarfs should behave more like large planets, with specific molecular characteristics and a variable lifetime dependent on their twin star or tidal neighbors. The relative fraction indicates that in our Milky Way galaxy, about 70% of all the stars are found in spectral class M with a surface temperature around 3,000 K and with the longest lifetime of all stars. Since phenomena have a tendency to repeat themselves on a cosmic scale, we expect the preponderance of class M to be universal, and since the most abundant element, hydrogen, sheds photons by the formation of neutral hydrogen around that temperature of 3,000 K, then this most important generation of free photons starts on its redshifting trip to the CMB.

Figure 5.1 Stars' energy output in terms of spectral class

The seventh column of table 5.1 shows the product of mass with relative fraction, and since luminosity, given by the Stefan-Boltzmann law, is proportional to radius and temperature and thus a power function of mass M, usually between M^2 and M^4, then we find that spectral class M is the most important. The eighth column is represented in figure 5.1 and gives the product of column 7 with the lifetime that corresponds to cosmic star energy output in terms of spectral class. We notice an important asymmetric peak for class M in the distribution, but a Gaussian-bell-shape graph would be more likely found if the study of stars included all EM waves equally and was not dazzled by visible light. It is the EM output from all stars and all spectral classes that are transformed and redshifted all the way to the cosmic microwave background.

5.1.1 Solar Radiation Spectrum

The blackbody spectrum of the sun with a temperature of 5,780 K is presented as an example of a star radiation spectrum in figure 5.2 with wavelength in nanometers on the horizontal scale and relative energy density on the vertical scale.

Figure 5.2 Sketch of solar radiation spectrum at sea level and above the atmosphere (Credit: Wikimedia commons)

A sketch of the typical and best-known spectrum of a star as seen from one of its planets is shown. That is the sun as seen by earthlings. The sun spectrum above the atmosphere is very close to a blackbody in the infrared part of the spectrum but with an excess in the visible portion because of stronger emission and a deficit in the ultraviolet because of absorption [42].

On the contrary, as seen at sea level with the sun at the zenith, absorption is present in almost the whole spectrum except in part of the infrared. It is most important in the UV and visible but with very pronounced bands in the infrared. About 5% of UV electromagnetic waves are transmitted compared with the 95% absorbed, whereas EM transmissions represent 45% in the visible and 50% in the infrared. In general, higher frequencies are more easily absorbed by atoms and molecules because they represent higher energies that react more easily according to the laws of quantum electrodynamics. UV radiation is absorbed by ozone (O_3), and other radiations are absorbed by water vapor, carbon dioxide, and the atmospheric gases.

In the next 10 billion years, the spectrum of the sun will gradually shift as nuclear fusion of hydrogen to helium increases to a maximum of 1.1 times its present value in about one billion years and then will shift to longer wavelength in the red over a four-billion-year period as the sun swells to the size of a red giant while burning hydrogen at an average rate of 600 million tons per second in its upper layers above the core. So after depleting about 15% of its hydrogen fuel in the next five billion years, the sun will start to fuse its helium core to produce carbon and oxygen in a short period of 100 million years and expel its outer layers to form an earth-sized white dwarf star surrounded by its planetary nebula shell expanding at an initial speed of less than 2,000 km/sec and gradually mixing with the surrounding clouds of star dust and molecular hydrogen to form new stars near the center of the galaxy, where the sun will have migrated down the spiral arm.

But the merger of the Milky Way with Andromeda in about four billion years may disrupt the sun's migration to the center of our galaxy. In the next five billion years or so, the star will slowly cool and become a black dwarf almost invisible against the cosmic background, and until about 10 to 15 billion years later, it is recycled with the merged galaxy in the crushing paws of the large next-door cluster and concentrated in the black hole of a newly born quasar.

5.1.2 Star Formation

If we want to understand the formation of galaxies and stars, we have to observe our own environment with all its curling eddies and whirling fluids and then transpose the model to the larger scales as the most likely scenario since similar processes are found at all scales.

Figure 5.3 Turbulent fluid flow with eddies and whirling cells

The model drawn to represent a turbulent fluid flow in figure 5.3 could apply to any kind of fluid, gaseous or liquid. The flow is introduced from the center left of the sketch and moving near the center toward the right. We note an enlargement of the flow to the right as the energy is dispersed in the fluid, but the most fascinating feature is the formation of eddies in directions opposite to the flow and all around the flow in 3-D with the resulting appearance in 2-D of clockwise eddies at the bottom and counterclockwise ones at the top. Some whirling cells will form when an eddy escapes from the main stream, and the cells will keep the same direction of rotation.

Stars are known to form in galactic molecular clouds from the available material containing a mass fraction of about 74% hydrogen, 24% helium, and 2% metals (all the other elements) corresponding to a mole fraction of 92% hydrogen, 7.5% helium, and 0.5% metals [63]. The fractions vary depending on the age and mixture of the local cloud within the more general galactic history. An older galaxy will concentrate a larger amount of dust and metals from its nucleosynthesis history and cannibal activities, while a younger one will mostly proceed to fuse helium from hydrogen although some dust and metal contamination is likely always present from prior maxibang and bubble cycles.

Thus, stars in those molecular clouds should emerge when the rotating eddies, with a conical depression or cylindrical hole at the center, are separated from the larger cloud to become an autonomous rotating entity gradually loosing kinetic energy until equilibrium is reached with gravity forces that will collapse the central mass and flatten out to a disk with its retinue of planets and satellites that have formed as smaller secondary eddies for the larger gaseous, low-density planets and gaseous to meteoroid contributions up to a stable orbit down the local spiral for the smaller satellites with a higher density. Whenever two primary eddies are close by, they may gravitate around a common barycenter and collapse to form one of the 35% of binary stars in the galaxy.

Molecular clouds vary in size from 100,000 to about six million solar masses. Their temperature varies between one and 100 K with an average evaluated at 10 K, which corresponds to a low molecular energy that will facilitate the gathering work of gravity over the masses involved in the formation of stars and planets. The particle density varies from 100,000 to 100 million per cubic meter and will locally reach more than one billion, especially away from the high-energy radiation and within the protected center of a cloud. Molecular clouds are also called star nurseries when they have the right characteristics to generate stars that may be observed in the making. Groups of proximal stars may form as open clusters with weak gravitational links, so some members may escape.

5.1.3 Main Sequence to Neutron Stars and Black Holes

The main sequence is the continuous band of stars that stands out on color-magnitude plots known as Hertzsprung-Russell diagrams [68]. The sequence represents all these newly born stars, also called dwarf stars, producing energy from the nuclear fusion of hydrogen to helium in their dense core at a temperature of at least 10 million K. These stars keep an equilibrium between the outward thermal pressure generated by fusion and the inward compressive pressure from gravity for a short period of time of less than one million years for massive stars with efficient radiation-convection energy transport and an extremely long period that could theoretically exceed one trillion years for wholly convective subdwarf stars. Slowly rotating or nonrotating main sequence stars with less than 1.4 solar mass, the Chandrasekhar limit, will form helium from the proton-proton complex chain reaction, and above that mass and 17 million K, the carbon-nitrogen-oxygen, or CNO cycle, is more efficient at producing helium from hydrogen.

With the exception of the long-lived subdwarfs, which may avoid a cataclysmic end for a long period of time, and stars with a mass less than the Chandrasekhar limit of 1.4 times that of the sun, some white dwarfs with a large companion and nonrotating stars larger than 1.4 times that of the sun may experience at least one explosive event in their lives. These events are called novae or supernovae depending on their mass and related history.

Novae are recruits from the 35% of binary stars in the Milky Way or the equivalent in other galaxies. Usually, the gas from a red giant is siphoned by tidal gravity by its white dwarf companion which builds up an accretion disk until the star reaches the Chandrasekhar limit, where fusion reactions occur in an abrupt manner between the core and the surface with a sudden rise in the luminosity of the white dwarf by factors as high as one million, followed by a gradual decrease. Any unprotected fusion reaction not far from the surface of a star may obviously be the source of high-energy X-rays or gamma rays.

Solitary stars like the sun and up to 1.4 times the mass of the sun may evolve as red giants when they start burning hydrogen in the layers above the core and later fuse helium to carbon-oxygen in the core. They may expel their outer layers at high wind speeds without the explosive death of supernovae. The remnant is a planetary nebula that has, at the center, a white dwarf with a carbon-oxygen core that will gradually radiate energy until it becomes a black dwarf.

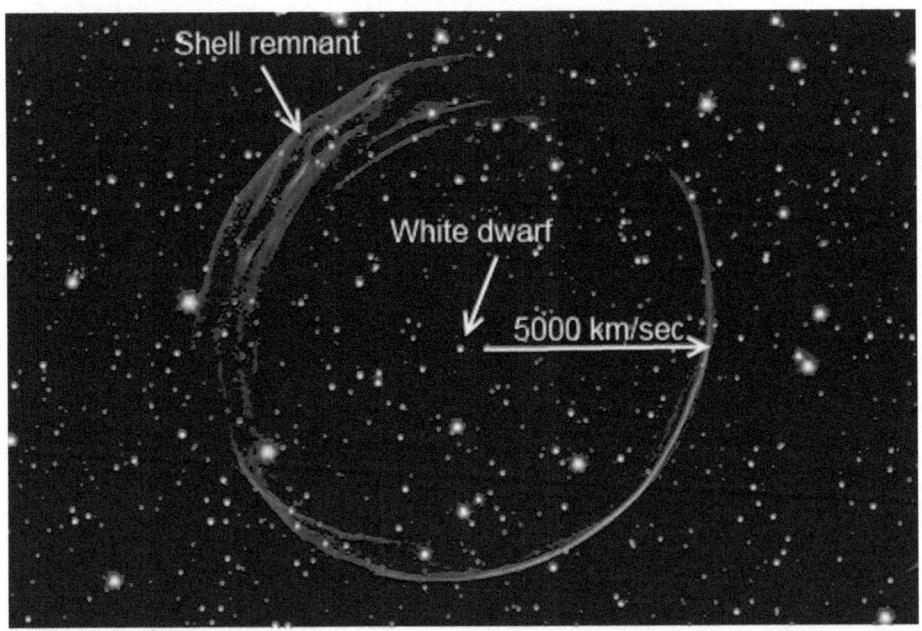

Figure 5.4 Transparent shell from supernova SNR 0509
(Credit: NASA Hubble Space Telescope)

Supernovae form the more cataclysmic class of stars in the family. Figure 5.4 shows a transparent shell remnant from supernova SNR 0509, also shown on the front cover of this book, taken by NASA's Hubble Space Telescope. A white dwarf or neutron star left over from the progenitor star is visible in the center, and an expansion velocity of 5,000 km/sec has been measured [70]. For the more energetic supernovae, the expansion velocity may reach 30,000 km/sec, which is 1/10 the speed of light. The amount of energy involved in the implosion and flash fusion of a few seconds represents 1×10^{44} joules for the lower masses and up to many times that quantity for giant stars. The luminosity increase may be many billion times that of the sun, making the supernova temporarily brighter than a whole galaxy [78].

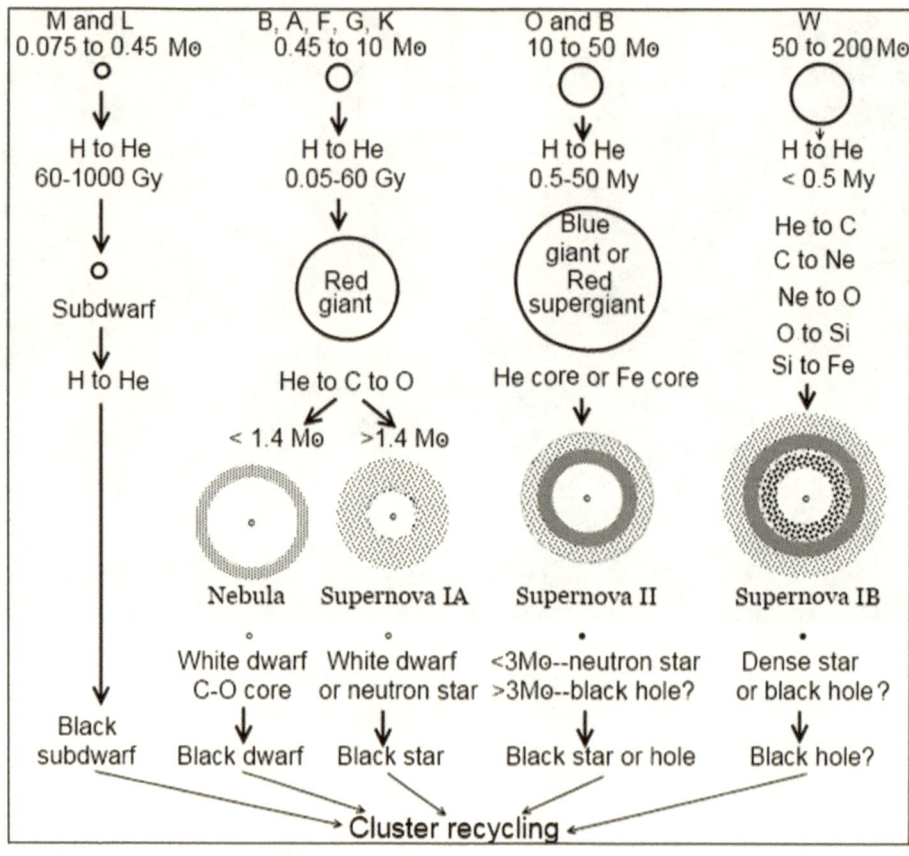

Figure 5.5 Curriculum vitae for different masses and classes of stars

Graphical curricula vitae for different masses and classes of stars are presented in figure 5.5, where the first row indicates the spectral class of the stars, followed in the second row by star masses in comparison to the sun's (M is for mass, and the round symbol with a dot in the middle represents the sun). The circles drawn are only sketches and are not proportional to real star sizes. The next two rows show that main-sequence stars start with hydrogen-to-helium fusion that will last for a number of megayears (My) or gigayears (Gy) depending on temperature and the efficiency of the fusion cycle. The bottom 2 rows indicate that no matter the status of the dead star, they are all eventually recycled and crunched in a local cluster in preparation for the next maxibang.

The formation of type Ia supernovae may be from accretion in a binary star system, where the larger star evolves more rapidly to the status of white dwarf and picks up material from the outer layers of its inflated red giant close companion until it reaches the Chandrasekhar critical mass, followed by rapid fusion implosion and an expanding shock wave producing the supernova explosion. This type of supernova may be considered as a standard candle to determine cosmic distances because of a characteristic light curve whose luminosity is produced by radioactive decay of nickel to iron. But this may be questioned since it assumes that the critical mass is always exactly the same for each explosion in type Ia. Some error allocation will probably have to be made for mass and surrounding variations, like a companion star or neighboring dust and clouds. Other scenarios to produce type Ia supernovae imply a merger of two white dwarfs or a merger of a white dwarf with any other star as long as the critical mass limit is reached. But it is easier and more natural to explain accretion of material down a spiral when a disk is present since it respects physical laws of merging.

Supernovae type Ib, Ic, and II are born from stars with masses of 10 solar masses and more that rapidly evolve from hydrogen burning to helium and metals up to the iron limit. When hydrogen is exhausted, the star maintains equilibrium by contracting under gravity and creates internal pressure against collapse by helium-to-carbon fusion at a temperature of about 100 million K. And when helium fusion terminates, then carbon fuses to neon at 600 million K, and the cycles of contraction-fusion continue at ever higher temperatures with neon, oxygen, and silicon until no more fusion can take place, and iron photodisintegration by gamma rays is reached around 6 billion K. The latter reaction consumes energy and will produce the central implosion and supernova.

The core may collapse at velocities of about ¼ the speed of light with a tremendous increase in temperature at 100 billion K and density that will produce a neutron core and a neutron star with the massive escape of neutrinos. The neutron star may also recoil from a companion star and thus be spinning and moving at a higher speed than the original star. For a progenitor star of less than 20 solar masses, further collapse is halted by the degeneracy pressure of the neutrons while the shock wave carries enough energy to expel the outer layers that will form the visible supernova remnant. Above that mass, and if no higher step in degeneracy pressure is present, like quark, electroweak, or any other quanta pressure, then a black hole may result. For stars with more than 50 solar masses, a lot of work is still ongoing for stability in the models of intermediate and final steps. The extreme heat and availability of neutrons may produce most of the elements beyond iron in type II supernovae. These elements from supernovae will mix with the other elements and become the major source of heavy elements that will enrich the molecular clouds and form basic molecules and dust grains before new star formation.

5.1.4 Metallicity and Redshift in Astronomy

The metallicity of stars, galaxies, or clouds represents the proportion of chemical elements other than hydrogen and helium that can be measured by spectroscopy [91]. The metallicity of the sun is 0.02, and its helium mass fraction is 0.25, while the hydrogen mass fraction is 0.73 for a total fraction of 1.00. Metallicity is usually expressed as the logarithm of the star's metal abundance compared with that of the sun; for example, the ratio iron to hydrogen (Fe/H) is frequently used. A star with a logarithm of +1 has 10 times the metallicity of the sun and 10 times less with -1.

The first stars are formed when a young bubble expands and cools down. They are born from the most pristine gas that contains only a small amount of residual metal dust. They have a spectrum showing a low metallicity and are the oldest stars in a star cluster or galaxy and are called Population II stars. On the contrary, Population I stars are younger and show a higher metallicity like our sun. In the big bang model, a Population III star was added in 1978 to account for massive virgin stars hypothetically born as the first stars after the big bang and prior to the explosion of any metal producing fusion stars and thus with hydrogen-to-helium fusion only and no metallicity, but no such stars have ever been discovered.

Studies of metallicity in nearby to high redshift galaxies, including damped Lyman alpha (DLA) systems, fail to demonstrate any evolution as predicted by the big bang model [92][93]. Large spreads of metallicities over two orders of magnitude are found for the same redshift. Except for a lower metallicity selection effect and poor sampling above z=2.5, those galaxies do not exhibit the cosmic chemical evolution from low metallicities at high redshifts to high metallicities in the local universe at low redshift.

This means, as expected in the multibang model, that the same physical processes of bubble expansion are in place in the local cosmos and at high redshifts wherein high galactic mass is generally associated with older galaxies and high metallicity, and high star formation rate is generally associated with younger galaxies and low metallicity. Note that all kinds of mixtures may be found in galactic mergers. For example, a low-surface-brightness or a high-surface-brightness galaxy may be temporarily enriched from the dusty influx of metal bearing tidal gas with the consequent rise in the star formation rate.

5.2 Families of Galaxies and Clusters

There are about 33 trillion galaxies in the cosmos, including an estimated 5% of void galaxies within the empty portion of the bubbles. They come in all kinds of size and shape. They are found isolated or in groups within the walls and in clusters between bubbles. The merging process is a major dynamic element in the formation of large galaxies, groups, clusters, and quasars.

Spiral galaxies, with their elongated arms forming flattened disks and bright central bulges often stretched in a bar shape, form the best-known and most outstanding family of galaxies. Elliptical galaxies, on the other hand, look like discrete stars or lone bulges with no spectacular shape. Irregular and colliding galaxies range from the extremely complex mixture of stars, clouds, and dust to simple irregular shapes. The spiral type of galaxy represents about 35% of the total within 30 million l-y of the Milky Way, while the elliptical family includes 15%, and the larger part of the pie goes to the irregular type, comprising around 50% [48].

Clusters are the locus toward which all nearby groups of galaxies, lone galaxies, and surrounding gas converge since the latter is driven by kinetic energy from expansion and by gravitational attraction. Smaller clusters are also moved by gravity and expansion energy toward the larger clusters. Thus, velocities are directly proportional to these two driving factors. Care must be taken not to interpret velocities at face value, but take all dynamic factors into consideration and thus avoid or circumvent the dark matter model.

Important research activities are currently focusing on what kind of material is present in galaxies and clusters and how it came about [69]. What are the limits, contributions, and exchanges between material phases such as ionized, neutral, and molecular? One hypothesis includes the presence of molecular hydrogen, including ¼ of the mass in helium, everywhere in the low-density voids, except in and around the filamentary structures associated with galaxy formation and clusters since that molecule is extremely stable and since EM waves may have a mean free path of many billion years in the rarefied medium between clusters and galactic walls. A more recent hypothesis is based on the analysis of quasar light that travels over large redshift z distances and in which specific absorption lines are attributed to the crossing of filamentary structure and where ionized hydrogen, including ¼ of the mass in ionized helium, would fill all the low-density voids and be invisible to EM waves since ionized atoms are a saturated plasma and nonreactive to high-energy radiation.

There are no definite answers yet, but many clues are under study in and around galactic walls and clusters. Meanwhile, to avoid the "chicken and the egg" priority problem, we may use the simplest hypothesis (Occam's razor) where we go from the original ionized atoms out of the bang to an expanding neutral hydrogen shell while preserving ionized atoms in the void within the bubble. Then molecular hydrogen will be formed directly and with dust as a catalyst within the expanding shell. This model is simpler than the alternate and more complex chain of events in the big bang model where we would go from the ionized phase after the bang to neutral hydrogen and then back to reionized hydrogen in an expanding universe where the latter must cool down to form neutral hydrogen again and then molecular hydrogen. The latter, H_2, is mostly expected to be formed, in the big bang, with the catalyst help of dust particles that would have been generated in a special unexplained process prior to the first star formation outburst [28].

So the formation of molecular hydrogen H_2 may proceed in the early phase of the expanding and cooling bubble shell after the formation of some neutral hydrogen at a temperature of about 3,000 K, with the following reaction $H + H^- \rightarrow H_2^- \rightarrow H_2 + e^-$ thus freeing an electron for the next cycle. Although there is competition for alternate reactions with the H^- anions, we know that the above reaction is probably the dominant one since the first stars are most likely formed from cold and dense molecular hydrogen clouds. When two atoms of hydrogen combine to form a H_2 molecule, then only ½ the number of particles is left, with a consequent drop in the kinetic energy and temperature of the system.

The formation of H_2 will be more important within the shell and preserved from destruction by high energy photons since the shell is swiping a volume where a small amount of dust was left over from the previous bubbles. This dust may be a shield and a useful catalyst to produce additional H_2 from neutral hydrogen. Thus, a picture emerges where we have an expanding shell made of molecular hydrogen clouds protected on both sides from high energy photons emitted by neighboring active bubbles by a layer of neutral hydrogen and also on both sides by a more exposed layer of ionized hydrogen. The inside layer of ionized hydrogen and helium within the shell would then leave behind the tenuous low-density material forming the large inside volume of the bubble.

5.2.1 Spiral Galaxies and Star Globular Clusters

The spiral family of galaxies is classified according to the size of the central bulge, where *Sa* includes the largest ones; *Sb*, the medium size; and *Sc*, the smallest central bulge [76]. Moreover, when a bar is present across the central bulge, letter *B* is inserted and yields the following symbols: *SBa*, *SBb*, and *SBc*. Generally speaking, spiral galaxies with up to 200,000 l-y in diameter exhibit two or more arms forming the disk, a central bulge with or without a bar, large gas and dust clouds, the generation of large-mass and short-lived *O* and *B* stars, older stars giving a reddish core and a reddish tint to the halo on both side of the disk, and young stars giving a bluish hue to the disk.

Spiral galaxies and star globular clusters seem to have a similar origin and a different life path. Both may be formed or rejuvenated when the shell of an expanding bubble crosses an old wall or, more generally, clashes head-on with the expanding shell of another bubble.

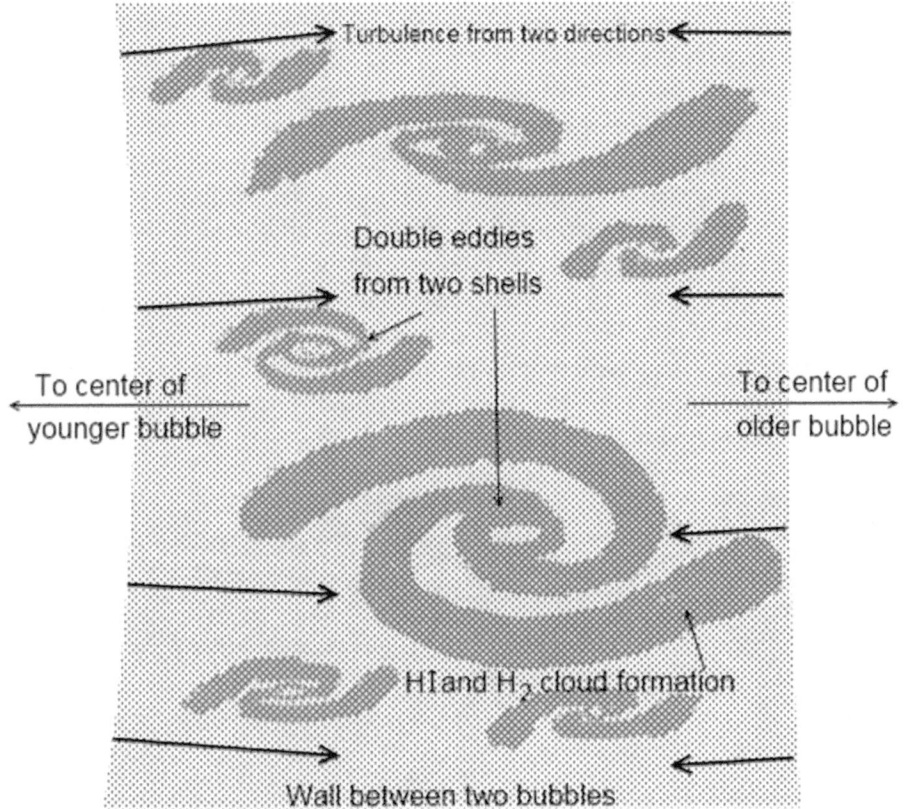

Figure 5.6 Galaxy formation between two expanding shells

Figure 5.6 is a blow-up view of the formation of a group of galaxies from two expanding shells of neutral hydrogen and molecular hydrogen, where the younger bubble with the strongest curvature is to the left and the older bubble to the right has a lower curvature indicated by arrows. In such a clash, differential turbulence is created from two almost opposite directions, while the density of the molecular gas, always including about ¼ helium by mass, increases from the contribution of the two shells and compaction across the common wall. Note that the low-density material within the huge internal volume of bubbles is likely an ionized and invisible gas escaped from the edge of the expanding shell and representing about 90% of cosmic volume. Since the bubbles are likely filled and lined with ionized gas, we should eventually observe some signals from hydrogen-and-helium emission spectra.

Eddies with a large range of sizes start their spiraling, spinning motion and flattening of an extended arm from the transfer of kinetic energy in the collision of the two shells. As the kinetic energy decreases, eddies will separate from the main molecular mass and generally form a more symmetric spiral with two main arms. At the center and along the arms in the disk, dense molecular hydrogen will locally spin and collapse under the force of gravity and be followed by massive star formation. Larger spiral structures will start on a first rampage mission to attract all nearby clouds and small companions and increase their elongating arms. The smallest ones will be the first to be stripped of their gaseous arms and be left with naked bulges that will become the cohort of globular star clusters and small elliptical and irregular galaxies usually accompanying the larger spiral galaxies. Midsize galaxies in the neighborhood will also be cannibalized but over a longer period of time, while star formation proceeds in protected high-density areas.

Figure 5.7 Comparison of a hurricane with a galaxy
(Credit: NASA/US NOAA and NASA/HST)

A comparison between a hurricane and a galaxy is presented in figure 5.7, where striking similarities and differences can be observed, like fractal patterns at different scales. On the left, hurricane Frances in the year 1996 photographed for U.S. National Oceanic and Atmospheric Administration by a NASA satellite above earth, and on the right, galaxy M51, also called the Whirlpool Galaxy taken by NASA's Hubble Space telescope. Similarities include the more massive look of the spiraling arms and their widening away from the center. The radius is about 500 km for the hurricane versus 43,000 l-y (4×10^{17} km) for the galaxy. Wind speed is around 200 km/hour in a hurricane versus rotation speed of about 200 km/sec (3,600 times more) in a large galaxy. But one interesting and questionable difference is the center where a dark hole appears in the hurricane while a bright spot is observed for the galaxy. The eye of the hurricane in the center is a relatively quiet and low-pressure location toward which winds converge. It is surrounded by walls forming a cylinder where cloud density and wind speed reach a maximum. On the other hand, the center of a galaxy is a massive high-density location toward which the surrounding masses converge by gravitational attraction. It forms a spheroidal bulge where the most central part is hidden by the bright light from the bulge. So what is hiding at the center of a galaxy? How about a black hole?

Figure 5.8 Moments of inertia from four geometrical shapes

In physics, the rotational kinetic energy about an axis is expressed with the following equation:

$$K = I\omega^2/2 \qquad (5.1),$$

where K is "kinetic energy," I is "moment of inertia," and ω represents the "angular velocity" about an axis. In figure 5.8, the moments of inertia of four different geometrical shapes are drawn and identified, from left to right, with decreasing fractional values, also expressed as the percentage of a fixed common mass and corresponding to the equation above each shape.

We notice that for a fixed mass, the moment of inertia and, consequently, the kinetic energy are highest for a hoop or cylindrical shape like the one we find at the center of a hurricane or at the center of any whirling fluid. A large drop to half the kinetic energy is found if the shape changes to a solid cylinder or a flat disk like that found in star formation with planets or in a spiral galaxy.

The next step-down in kinetic energy is the solid sphere where only 40% of the energy is left, like in stars, planets, and the spheroidal bulges in galaxies. Finally, when the length of a bar L is set equal to twice the radius, then only 33% of the kinetic energy would remain for a bar-shaped structure like the one found across the center of barred galaxies. The main idea here is that not only the velocity but also the shape of a structure may be modified by kinetic energy variations.

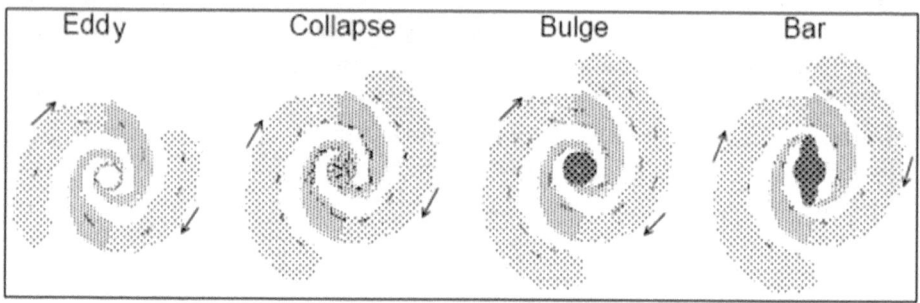

Figure 5.9 Spiral galaxy evolution from eddy to barred system

In figure 5.9, we transpose this physical information to explain the evolution of a spiral galaxy, where the collision of two shells produces an eddy with a cylindrical or ring-shaped core that will collapse to a filled cylindrical core under the pull of gravity as kinetic energy decreases and while a disk is generated from the flattening of the arms. An initial burst of star formation will produce metals from fusion and inject it more or less explosively into the arms of the galaxy to the benefit of the next generation of stars. The arms will increase in length and mass from the input of neighboring gas clouds and smaller galaxies.

Notice that the stars, dust, and gas clouds are gradually moving in the arms toward the center of the spiral as shown by the arrows in a way similar to the clouds in the arms of a hurricane spiraling toward the central low-pressure eye. The latter movement in galaxies funnels and concentrates all the gas and dust material in the narrower portions of the arms until density and temperature in the molecular clouds are sufficient to produce new stars.

As kinetic energy keeps going down, the center of the galaxy is transformed into a larger spheroidal bulge mostly from the accumulation of stars from the first generation while the arms produce stars like our solar system and stars destined to a long life. The production of explosive stars continues at a lower and steady rate in the giant molecular clouds. Accretion of material from the surroundings is prompted by the gradual strengthening of the gravitational field in proportion to the increase in galactic mass.

Finally, as kinetic energy decreases near the center toward the lower limit associated with a bar, the velocity of the inward-spiraling material in the arms decreases to the one around the bulge, so a bar will gradually grow across the central bulge until an SB galaxy emerges. Thereafter, the galaxy will pursue its dog-eat-dog activities with the neighbors until it forms a lenticular galaxy that will be pushed and pulled into a cluster where, in turn, it will be gobbled up by a giant elliptical galaxy.

5.2.2 Lenticular and Elliptical Galaxies

Lenticular galaxies are classified as spiral-elliptical *S0* or as barred spiral-elliptical *SB0* between the most spherical elliptical ones in E7 and the largest spirals in Sa. In other words, an S0 lenticular galaxy like NGC 5866 or the large Sa Sombrero Galaxy (M104) present mixed hybrid characteristics between spiral and elliptical galaxies since a disk surrounded by an elongated or oblate spheroid of bright light is observed. Thus, large barred spiral galaxies may exhaust their supply of star-forming gas clouds during their trip toward a cluster over a period of many billion years with the ensuing consequence that their arms will also be consumed and disappear, leaving only an elongated spheroid. As the mass and density of the spheroid increases with mergers, we should observe an increase in kinetic energy that will transform the galaxy into a giant elliptical with the aspect of a rounded sphere.

The main characteristics of elliptical galaxies are the presence of a reddish color produced by old stars and the absence of gaseous and dusty arms, thus hindering new *O*- and *B*-type star formation. The shapes vary from the sphere, designed type *E0*, to the most elongated cigar-shaped, designed *E7*. Giant elliptical galaxies reaching a few million light-years in diameters are by far the largest, but like the size of fishes in the ocean, they are outnumbered by a factor of 10 by dwarf elliptical galaxies.

Elliptical galaxies may be divided into two main groups: the first group involves mostly younger and smaller galaxies that have lost most of their gaseous clouds to the profit of more massive neighbors, and the second group concerns generally older and larger galaxies that have exhausted most of their star-forming clouds.

All galaxies, including the elliptical ones, are expected to hide in their center a black hole proportional in size to the mass of the galaxy so that accretion of new material should also feed the black hole. Larger elliptical galaxies are so dazzling that we cannot observe their intimate core and activities, but recent X-ray and radio wavelength data indicates that they hide one central black hole and often numerous secondary black holes from merged galaxies and also from globular clusters or high-energy activities in massive stars or binary star systems.

5.2.3 Colliding and Irregular Galaxies

Irregular galaxies are characterized by the absence of a clear geometrical form where no complete ring or spiral arms may be observed. These oddities may show asymmetric or numerous bulges from mergers. They are generally rich in bright O and B young stars born from their molecular clouds.

The smaller and more numerous irregular galaxies are satellites to larger galaxies, like the two Magellanic Cloud galaxies that are gravitationally bound to our Milky Way galaxy; tidal effects produce partial or total deformation of these satellite galaxies, which end up looking like a bad dream. The larger irregular ones are the result of one or many encounters with fair- to large-size galaxies where a merger may take place or where the kinetic energy of one galaxy may sometimes, like a pendulum, be sufficient to swing back and forth in the gravitational field of the larger galaxy. Others with sufficient kinetic energy may escape the field, with damage mostly limited to the capture of gas and dust by the more massive galaxy, but in any case, a common barycenter and many secondary barycenters may be present and dictate the reshaping of the whole material. Many theoretical mathematical models have been developed on star and galactic dynamics and presented by authors such as Binney and Tremaine [4].

5.2.4 Active Galactic Nucleus or AGN Galaxies

The black holes at the center of massive stars and of all galaxies are dynamic locations where high velocities and a strong magnetic field are present and produce many different phenomena. There are recurrent time-limited EM signals associated with large galaxies. They vary from low-frequency radio waves to high-frequency X-ray and gamma ray signals.

One class of galaxies with active nucleus is the Seyfert galaxies, which are generally found in larger spiral galaxies like NGC 1068. They show a bright compact nucleus that may imitate a quasar in visible light since the arms are occulted by the bright light. Seyfert nuclei will radiate strongly in infrared, optical, ultraviolet, X-ray wavelengths, and for about 5% of them, also in radio wavelengths. Broad emission lines very close to the center indicate star, gas, and dust velocities up to 5,000 km/sec around the compact nucleus, which is most likely a black hole.

Radio galaxies are mostly found in elliptical galaxies, like M87 at the center of the Virgo cluster, where they radiate more strongly at radio wavelength than spiral galaxies. They present two emitting radio lobes located around the apex of two high-velocity and opposite jets composed of energetic particles like electrons moving along helicoidal curves in a rotating magnetic field. The relativistic high-velocity electrons will first emit X-ray radiation along the jets and then emit lower-frequency radiation with typical strong radio wavelengths in the extensive lobes.

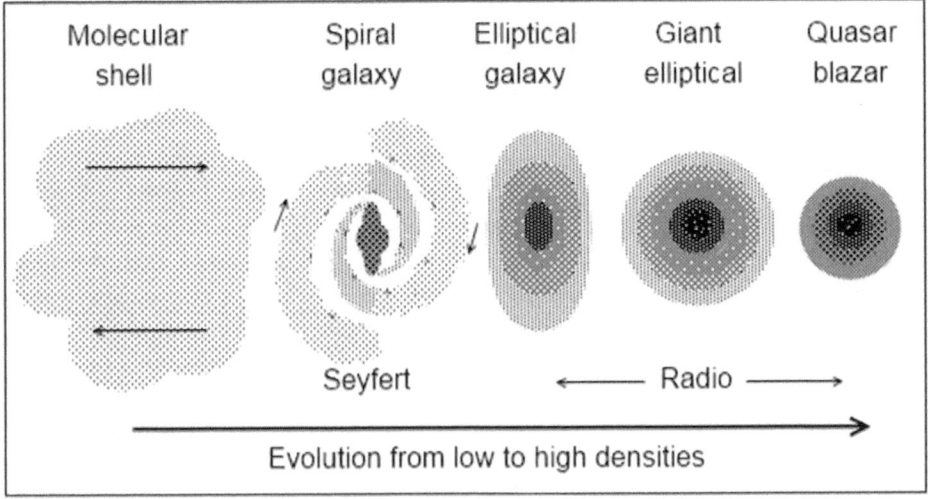

Figure 5.10 Galactic evolution from low to high density

Galactic evolution is summarized in figure 5.10, where galactic types progress in density from left to right. We start with a molecular shell expanding and clashing with a neighboring shell, where an eddy produces a spiral galaxy that evolves to larger size and to a barred spiral where strong EM emissions from infrared to X-ray may indicate a Seyfert galaxy with an active galactic nucleus.

Then spiral arms are gradually consumed in the intermediate step of a lenticular galaxy before emerging as an elongated elliptical galaxy where radio emission is often a witness of strong activity in the galactic nucleus, including two high-energy opposite jets and lobes. Large elliptical galaxies keep on feeding on the cluster's smaller subjects until they become giant ellipticals with periods of strong EM emission from the dense and very compact nucleus with a massive black hole at the center. Finally, when approximately one thousand galaxies and the associated gas representing around six times the galactic star mass have been cannibalized by a central super giant galaxy, we end up with a quasar or a blazar depending on the orientation of the giant mass and its nucleus [53].

The hypothesis of a black hole at the center with infinite density is unacceptable from first principles like conservation of energy and space-time concepts. We know that large masses are present in massive stars or accumulate in the core of galaxies, but these are most likely stopped in their gravitational contraction at different steps as suggested by quark star models and other anticipated but not proven subquark quantum mass energy compaction schemes or by the presence of some superluminal black energy. With this in mind, a black hole is an extremely dense mass located at the center of a star or galaxy from which no light or EM wave can escape but with no influence on the gravitational field other than being stronger and more concentrated. The latter field escapes unscathed since gravity is not limited by the speed of light.

5.2.5 Intra- and Perigalactic Clouds and Dust

With the extreme galactic variation in size, location, and dynamic interactions as presented in hundreds of written articles and images every year, we will only summarize some of the galactic characteristics and the extent of gas clouds and dust in and around solitary field galaxies and groups of galaxies located in the walls between bubbles [9]. For example, Andromeda and the Milky Way with their cohort of smaller galaxies may be considered as a group.

Cosmic galactic group parameters:			
Physical parameter	Number	Units	Remarks
number of groups	6.9848E+11	groups	if 200groups/bubble
average radius	1.50	Mpc	megaparsec
average radius	4.89E+06	l-y	light-years
average radius	4.6308E+22	m	
average volume	4.1597E+68	m^3	cubic meters
average mass	1.39E+42	kg	
average mass	8.32E+68	Mp	proton masses
average density	2.00	Mp/ m^3	proton mass/cubic meter
nb large galaxies	2	galaxies	1.6E+11 stars each
nb small galaxies	38	galaxies	1.0E+10 stars each
nb galaxies/group	40	galaxies	2 to 50 galaxies
nb of stars/group	6.98E+11	stars	sun masses
nb of stars/galaxy	1.75E+10	stars	sun masses
mass of stars/galaxy	3.47E+40	kg (sun masses)	intragas included
mass of stars/group	1.39E+42	kg	sun masses

Table 5.2 Cosmic galactic group parameters

In the cosmic parameters for galactic groups in table 5.2, we assume a ratio of two hundred groups per bubble or 17 groups per wall, which is an average since each bubble shares common walls with up to 12 neighbors according to its past history of expansion and encounters. The average spherical radius is only indicative since most groups are expected to inhabit generally flattened 2-D walls. Our own group serves as the main model for the number and distribution of large and small galaxies, and our sun serves as the model for star mass although we know that red dwarf M stars are by far the most numerous with an average mass about ⅓ of that of the sun. So we find that galactic groups have an average density of two proton masses per cubic meter which is about 10 times more than the cosmic average. The number of stars per galaxy is the equivalent of a low average value of 17.5 billion solar masses since most galaxies are relatively small compared to large galaxies, such as the Milky Way with 160 billion solar masses.

Figure 5.11 Inverted black-and-white galaxy M 83 and surroundings as seen from a range of wavelengths (Credit: NASA/JPL-Caltech/VLA/MPIA)

When a galaxy is observed with a large range of wavelengths, a whole new world is discovered. For example, in figure 5.11, the Southern Pinwheel galaxy, M83, is a barred spiral located in the Hydra constellation at a distance of 15 M l-y which shows wide and long arms of neutral hydrogen when observed at the 21 cm wavelength with a radio telescope. These extraneous arms, not visible at optical or ultraviolet wavelengths, spiral and extend four to six galactic radiuses all around and away from the "darker" central visible part of the disk of M83 where we can recognize the arms, ellipse, and bar of the galaxy. They can be compared to unspooling fluffy ribbons originating from the torn-apart neighboring galaxies in the process of being cannibalized by M83 and from the important residual gas left over from the two original shells that gave birth to the cradle wall. In the ultraviolet-wavelength telescope, these elongated arms look like a galactic skeleton since the energetic photons are emitted by young stars born out of the limited molecular clouds concentrated within the thick protective shield of neutral hydrogen and dust. A similar phenomenon is present to a lesser extent around Andromeda M31, around our Milky Way galaxy, and more or less around all cannibalizing galaxies.

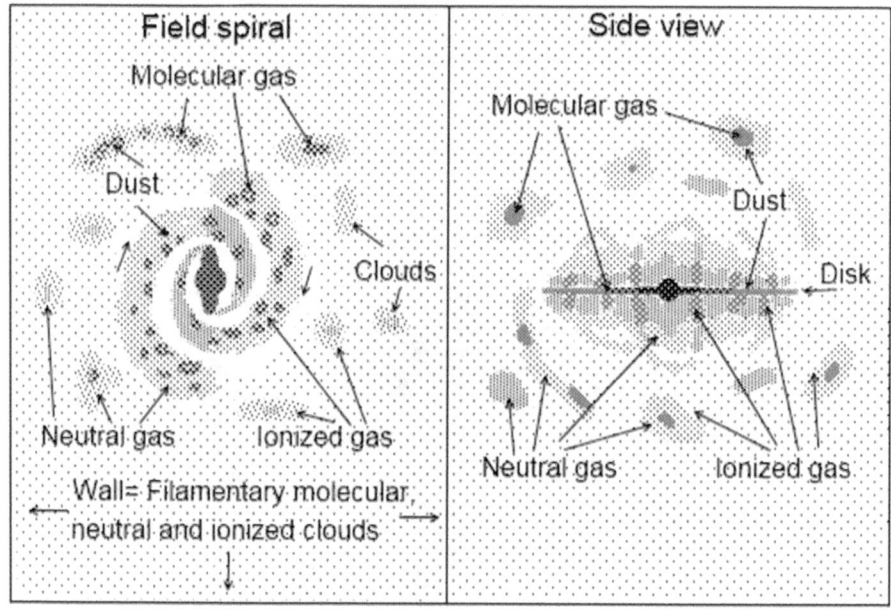

Figure 5.12 Field spiral and surroundings in a wall

The drawn sketch represents a barred spiral galaxy, which is typical of large galaxies located in a squeezed wall between bubbles. In figure 5.12, we show a face view on the left and a side view of the galaxy to the right. The wall, mostly made on both sides of ionized gas from two shells exposed to high-energy photons, includes a mixture of neutral gas and molecular hydrogen where the gas is protected from ultraviolet and high-energy photons toward the center. It also includes dust where dense, hot stars have released their metals within a cloud.

The arms of the galaxy are the location where star formation takes place from the giant molecular clouds that are gradually funneled and spiraling in toward the center. It is also the site where supernovae expulse at high velocities their external layers containing most metals that will combine to form dust and larger molecular particles. Part of this material involved in the implosion/explosion reach a sufficient velocity to escape from the central plane of the galaxy and reach a distance of up to a few kiloparsecs and then either escape from the galactic gravitational field or slow down and fall back like a fountain.

A lot of activity is associated with the end of a bar, when present, such as intense star formation where stars are integrated to the particular kinetic energy and velocity of the bar and where an important outward wind may blow particles away from the high star density and form clouds above the disk. The bulge usually contains older, more mature stars and very little molecular gas, so star formation is halted near the galactic center, but a large amount of ionized gas from the stars is present and steadily accreted to the central black hole [5].

As we move away from the nearby galactic corona for which most of the material may come from the galaxy itself, we find clouds and filaments, as in the Southern Pinwheel Galaxy M83, that have been pulled from neighboring and usually smaller galaxies [34]. This gas and dust material may contain an external envelope of gas ionized by high-energy photons originating from the galaxy. The interior of these stolen clouds may contain protected molecular hydrogen surrounded by neutral hydrogen and dust. In merging situations, a filament of dense and cold 10 K or so molecular hydrogen may produce by compression a series of starbursts, visible in ultraviolet and sometimes in optical observations, during its trip to the more massive galaxy.

5.2.6 Clusters of Galaxies and Intercluster Medium

As explained previously, galaxies are born and generally lined up in the plane of the wall formed by two colliding shells from two expanding bubbles. While they are gradually displaced sideways across the wall by the kinetic energy of the expanding bubble, they are transformed to cannibalizing spirals that will grow in size, mass, and density until they move out of the wall into the concave triangular section toward the closest tetrahedral cluster. Then they will mute to S0 and to elliptical galaxies since they are cut off from their fresh supply of cold and dense gas. The next mutation step brings them in the stomach of a giant elliptical galaxy with some of the surrounding hot compressed gas falling in. Finally, a quasar will form in a large hexahedral cluster.

Cosmic cluster parameters:			
Physical parameter	Number	Units	Remarks
number of clusters	6.9848E+09	average clusters	500 galaxies/cluster
average radius	4.10	Mpc	1.5 to 10 Mpc
average radius	1.34E+07	l-y	light-years
average radius	1.2658E+23	m	
average volume	8.4946E+69	m^3	cubic meters
average mass	1.70E+44	kg	
average mass	1.02E+71	Mp	proton masses
average density	12.00	Mp/m^3	proton mass/cubic meter
nb large galaxies	50	galaxies	1.6E+11 stars each
nb small galaxies	450	galaxies	1.0E+10 stars each
nb galaxies/cluster	500	galaxies	50 to 1000 galaxies
nb of stars/galaxy	2.50E+10	stars	sun masses
mass of sun	1.99E+30	kg	
star mass/galaxy	4.98E+40	kg	average
star mass/cluster	2.49E+43	kg	
gas mass/cluster	1.45E+44	kg	
ratio gas/star mass	5.84		about 6X more gas
Parameters for invisible gas in walls:			
wall gas volume	6.90E+80	m^3	out of groups + clusters
wall gas mass	1.12E+54	kg	
wall gas mass	6.73E+80	Mp	proton masses
wall gas density	0.97	Mp/m^3	proton mass/cubic meter

Table 5.3 Average physical parameters for cosmic clusters and gas in walls

The upper part in table 5.3 is showing cosmic-scale parameters for the average-size cluster, which contains about five hundred galaxies. Note that we have for each bubble the equivalent of two average-size tetrahedral clusters. After mergers, this will yield the large hexahedral-size cluster, including the mass of one thousand galaxies plus the equivalent gas mass of six thousand galaxies that will be compacted and produce the next bubble cycle. The average mass of one cluster is about 1.65 times the mass of one bubble, but its volume is more than three hundred times less, so cluster density is more than five hundred times that of bubbles. A cluster ratio of six times more gas than star mass has been applied in order to account for the large amount of high-energy gas detected within clusters. The last four lines of table 5.3 show the parameters for the generally invisible gas present in all the walls between active bubbles and outside of group and cluster volumes. This is not related to fossil walls within bubbles. The volumetric density of this gas is about five times more than the average cosmic density.

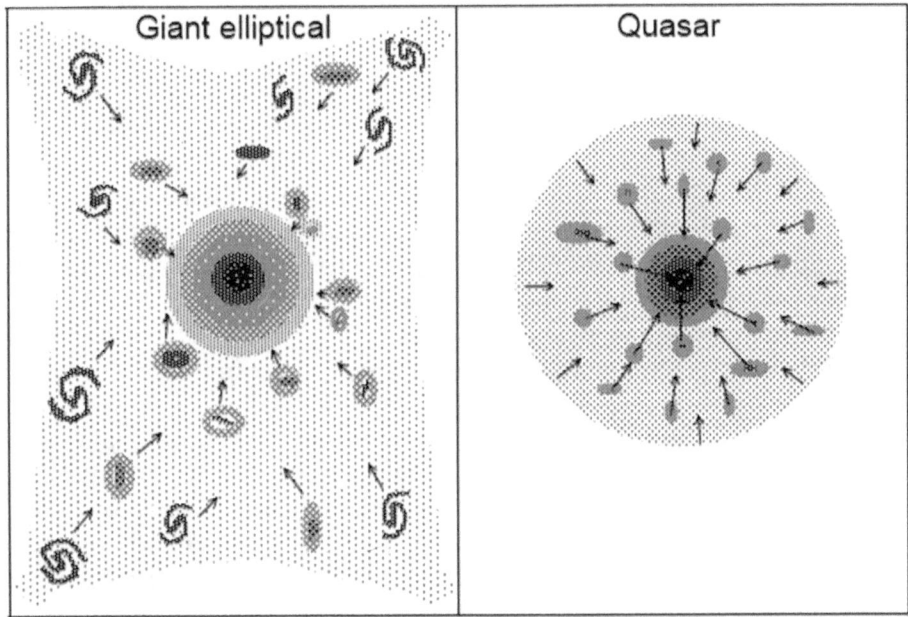

Figure 5.13 Cluster with a giant elliptical galaxy and a quasar

In figure 5.13, we illustrate a giant elliptical galaxy on the left and a quasar on the right, both located within a cluster of galaxies. The giant elliptical is spherical since it involves a very large mass with increasing density toward the center, like in a star or large planet. Spiral and elliptical galaxies formed in the walls are moving in the cluster via the concave triangular section between three bubbles, with all their surrounding gas, dust, and debris, including stars, planets, and satellites. The latter material is compressed and heated to a high temperature while galaxies are merging and falling in toward the central ogre from which a double jet of particles is emerging at high velocity and signaling the presence of an insatiable central black hole.

The quasar is the last and final phase in galactic evolution when the equivalent of a critical mass between five and 10 thousand galaxies has been gathered in a giant hexahedral cluster, including mostly hot gas and the smaller contribution of about one thousand galaxies. The cluster is gradually compressed and partly cooled by the emission of a large range of EM waves carrying some heat away. The last galaxies and the rest of the gas are rushed in toward the extreme black hole in the center over a period of a few hundred million years. In the end, like the phoenix consumed in a fire and reborn from its ashes, the quasar brings the whole cluster to a final contraction before its rebirth in a maxibang and expansion as a new local bubble.

5.3   Quasar Distribution and Dynamics

When we observe galaxies or quasars, the total number available for observation increases approximately in proportion to the distance up to a limit where they start to decrease because absorption, extinction, and masking by intervening walls, clouds, and galaxies gradually hamper their visibility in EM wave recordings by various land-based or orbital telescopes.

With the expansion of the shell of every bubble comes the clash with other shells and the formation of stars and galaxies where heat is produced and released as EM waves ranging from gamma rays to radio.

Quasars and blazars, depending on their orientation, are known to emit strongly over a large spectrum of frequencies [73]. They are likely the source of acceleration for some of the most extreme high-energy protons and helium-core cosmic rays. Two opposite and powerful jets of charged particles, like electrons and protons, are observed from the active black hole center of a large number of quasars. These jets represent the visible part of the particles accelerated to relativistic speed in the gyrating magnetic field of the rotating hooplike central mass from which nonthermal X-rays, ultraviolet, and optical photons are emitted in a way similar to synchrotron emission in a lab. At the end of the jets, a plume or a cloud of EM waves ranging from visible to radio is observed. Digesting all these galaxies and gas entrains some spectacular fountains of light since the energy output of a quasar is comparable to the luminosity of hundreds and even thousands of galaxies.

5.3.1 Quasar Distribution

We will first review the distribution of known quasars in the surrounding cosmos circa 2010, and then we will extrapolate the numbers to the estimated volume of the cosmos.

Figure 5.14 Distribution of known quasars from the Vizier catalogue

When we take the Vizier catalogue, the Schneider 2010 SDSS quasar catalogue where 105,842 quasars are presented, and plot the redshifts within interval increments of $z=0.25$ versus the number of quasars, we obtain figure 5.14. For the first two values, the number of quasars increases very rapidly and is followed by a steady but less important increase up to $z=1.75$, where a maximum is reached in the interval with about fifteen thousand quasars. Then a rapid decrease is observed up to $z=5.5$, where redshift will displace photons to hidden parts of the spectrum, extinction will mask them, and absorption will recycle a portion of EM waves coming to us from quasars located afar. A little kink is seen around $z=3$ and is probably related to irregular volume coverage.

Since we expect an average cosmic radius in the order of $z=10$ and since absorption and redshifts are the main handicaps, we should be able to record the presence of quasars up to that redshift by observation of the whole infrared to radio range of frequencies with more sensitive and accurate instruments than those presently available.

Absorption of ultraviolet photons by neutral hydrogen across the Lyman-alpha forest and of those with relatively high-energy by a variety of absorbers present in the expanding shells, walls, and clouds located between us and the quasars is a gradual phenomenon increasing with the number and intensity of encounters along the line of sight until no photons are left to be absorbed [67]. One neutral hydrogen atom in 10 thousand in a cloud is enough to produce complete Lyman-alpha absorption.

This gradual absorption and reddening of photons to a value of zero is expected in the multibang model, but it has been named the Gunn-Peterson effect and trough in the big bang model. In the latter model, it is based on an elusive reionization concept that never occurred. In 1965, Gunn and Peterson were assuming that most hydrogen atoms are the neutral kind in the cosmos. They never found the expected trough, blueward of the Lyman-alpha emission peak, in the spectral data from quasars at z=3 and came to the conclusion that most of the intergalactic hydrogen is ionized hydrogen. The latter conclusion confirms our multibang model where expanding bubbles are mostly filled with ionized hydrogen and helium. Further, a study by Hernquist and Springel (2003) indicates that photoionization from the stars, which are the main cosmic source of high energy ionizing photons, is constant over a large range of redshifts [85], thus rejecting any big bang reionization model.

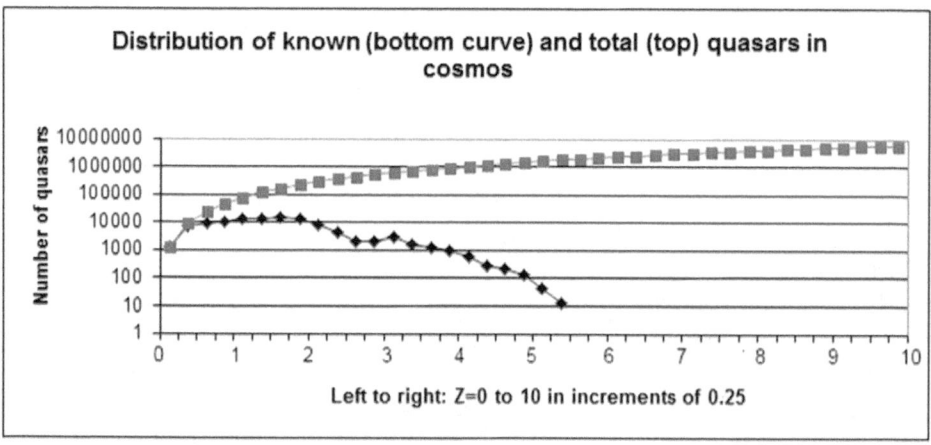

Figure 5.15 Known and projected distribution of quasars

The bottom curve of figure 5.15 is a repeat of figure 5.14 at a logarithmic scale, and the top curve is a volumetric projection in increments of z=0.25 of the total number of quasars that we may expect to eventually discover with improved instruments and time. Of course, the projection is an initial estimate where the last interval of z=0.25 contains over six million quasars. The constant total cumulative number at any epoch would then be over 83 million quasars in the cosmos, so the number discovered (about 105,000 and growing) represents only about 0.13% of the total. Also, note that these are spherical projections to obtain a total value although we are most likely not located at the center of such a sphere.

5.3.2 Quasar Lifetime and Dynamics

A table summarizing quasar parameters with their maxibang timeframe dynamics is presented.

Quasar and maxibang parameters			
Physical parameter	Number	Units	Remarks
mass of sun	1.99E+30	Kg	
number of quasars	8.30E+07	quasars	projected
bubble/quasar	4.2077E+01	ratio	
ave bubble radius	8.6177E+20	Km	
ave expansion vel	4,440	km/sec	across a bubble
ave radial exp vel	2,220	km/sec	
radial expan time	3.8818E+17	seconds	
ave bubble lifetime	1.2301E+10	years	
ave quasar lifetime	2.9235E+08	years	
quasar birth/death	3.52	years	average
quasars discovered	1.05E+05	quasars	Vizier/Schneider 2010
ratio total/discov	7.90E+02	ratio	for quasars
probab to observe	1 in 2,780 years	probability	for discovered quasars
nb clusters in bang	1	cluster	1 large cluster/bubble
nb galaxies in bang	1,000	galaxies	1 hexahedron with gas
ave mass in bang	3.40E+44	kg	1 large cluster
sun masses in bang	1.71E+14	solar masses	or 171 trillion suns

Table 5.4 Parameters for quasars and maxibangs

With the projected number of quasars and the ratio bubble/quasar in table 5.4, we find the average bubble lifetime from the expansion velocity and also the average quasar lifetime of about 300 million years, which implies that a quasar is born in the whole cosmos about every 3.5 years, and that one implodes to a maxibang in that same period of time. But since only one out of eight hundred quasars have been discovered, our chance to observe a new quasar or a maxibang is about one in 2,800 years if we keep observing that same known volume for that period of time. Therefore, to observe the birth or the death of a quasar in a human lifetime would require the steady observation of a much larger cosmic volume. The staggering amount of 171 trillion solar masses equivalent is involved in the triggering of a maxibang event. It includes the mass of one thousand galaxies and six times as much in gas contribution.

5.3.3 Quasar Emission and Absorption Spectra

Quasars are the strongest cosmic dynamos emitting EM waves over the largest spectrum, and their traveling redshifted photons must cross the equivalent of a steeplechase to reach us from their distant locations. Recorded photons from a quasar are displayed as a spectrum in terms of digitized wavelength versus flux density. It represents the relative number of photons that have been emitted by the quasar at the recorded wavelengths and part of those absorbed or missing within the same range. The redshifted wavelength is found at (1+z) times the rest wavelength so that for a redshift of one, for example, we find twice the rest wavelength.

Figure 5.16 Quasar Q1323+6530 at z=1.618 and some attributes [86]

The recorded portion of the spectrum of quasar Q1323+6530 at a redshift $z=1.618$ is presented in figure 5.16 with some attributes. The horizontal scale is expressed in wavelengths divided in Angstroms (1 Angstrom=1Å=0.1 nanometer) and runs from 2230Å at upper left to 3280Å at lower right. The vertical scale is the photon density coming from the quasar and expressed as a flux in erg cm^{-2} sec^{-1} Å^{-1}. The blueward and redward directions in the spectrum are indicated with arrows and mean toward shorter and toward longer wavelengths, respectively. Blueward of the Lyman alpha emission peak, absorption troughs are enhanced with a dark shade. Note that absorption troughs are present between 2,800 and 3,000Å although with a lower amplitude.

The redshifted spectrum location and residual emission peak of Lyman alpha, beta, gamma, and limit is shown with an arrow and followed by the rest wavelength as emitted by the quasar. Also, below each location, the energy in electron volt of the quantum jump is indicated for the rest wavelength emitted from the source at z=0 and also, in parentheses, we find the energy for the redshifted wavelength at $z=1.618$. For example, the Lyman alpha photon was emitted with an energy of 10.2 eV at the rest wavelength of 1216Å and is reduced to 3.9 eV after being redshifted to $z=1.618$, while crossing about 180 bubbles and reaching us at a distance of about 23 billion l-y. But note that even if the decreasing and redshifting photon energy from 10.2 to 3.9 eV cannot be absorbed by an electron in the larger quantum jump from level 1 to level 2 in the most abundant neutral hydrogen, it can still be absorbed in the 180 walls made up of clouds of gas and dust along the line of sight by a lower energy jump within hydrogen, helium, or metals.

Although the Lyman alpha emission should be a narrow line at the rest wavelength of 1216Å (redshifted to 3183Å on the figure), we observe a spread of emitted values on both sides of the peak at 3183Å because of the Doppler effect from particles emitted at high speed by the opposite jets of the quasar. This applies to all emitters moving away from us and thus producing a redward emission from the rest wavelength and also to those moving toward us and producing a blueward emission spread.

The amplitude of the spectrum and the number of absorption troughs are generally higher blueward from the Lyman alpha peak because of the preferential absorption of the Lyman energy between 10.2 and 13.6 eV by hydrogen atoms which represent 92% of all atoms. It is lower in the redward direction because there are less absorbers available to react to energies below 10.2 eV. Blueward of the Lyman limit at 2388Å, the spectrum shows only some noise and is flat near zero because all those photons with energies higher than 13.6 eV or lower than the rest wavelength at 912Å have been absorbed near the quasar by the intragalactic medium. The Milky Way, for example, is opaque to the extreme ultraviolet ionizing radiation in the range 912 to 62Å.

Figure 5.17 Partial spectrum from quasar Q0014+813 and vertical compaction [87]

The partial spectrum of quasar Q0014+813 at a redshift z=3.381 is presented at the bottom of figure 5.17 and at the top with a vertical compaction. The horizontal scale is in wavelengths divided in Å and runs from 5000Å on the left to 5450Å on the right. The vertical scale is the relative flux. The Lyman alpha emission peak is indicated with an arrow, and the blueward absorption troughs are named the Lyman forest because of the apparent density and larger amplitude of the troughs.

But when the Lyman forest is compacted by a factor of seven as shown on the top of the figure, then the apparent forest disappears. This demonstrates how scaling can change the aspect of a spectrum and suggests an interpretation. We observe that the Lyman alpha emission-high looks more like a hill instead of a mountain because of absorptions, and that increases the relative amplitude of the absorption troughs.

Figure 5.18 Partial spectrum from quasar Q0003+1553 stretched to the scale at the bottom of figure 5.17 [86]

A small part of the spectrum of quasar Q0003+1553 at a redshift z=0.45 is presented in figure 5.18. The horizontal scale is expressed in wavelengths divided in Å and runs from 1380Å on the left to 1460Å on the right. The vertical scale is a relative flux. Both horizontal and vertical scales have been stretched to the comparable redshift and amplitude of the Lyman forest at the bottom of figure 5.17. The Lyman alpha emission peak is on the right and outside the figure.

Thus, we observe that a quasar at z=0.45 may exhibit a very similar Lyman forest to that from a much more redshifted quasar at z=3.381 when similar scaling is applied. In fact, the photons from a quasar need only to be redshifted to z=0.33 to fall outside of the 10.2 to 13.6 eV main Lyman absorption range by neutral hydrogen. So our quasar at z=0.45 already contains all the main Lyman absorptions for the largest quantum jumps. But spectrum recordings from quasars vary with the quality and location of the field instruments and from the complex path of the photons.

Figure 5.19 Sketch of the average spectrum of a quasar around z=6 with some attributes

A handmade sketch of the average spectrum of a quasar at a redshift around z=6 is presented in figure 5.19 with some of the expected attributes. The horizontal scale is in wavelengths divided in Å and runs from 5500Å on the left to 9800Å on the right. The vertical scale is the relative flux. The Lyman alpha, beta, and limit emissions are indicated with an arrow. The surviving Lyman alpha emission shows a relatively high peak that makes the Lyman forest look like a decimated forest because of the relative amplitude and proportionately high absorption and reddening at such a high redshift. Blueward of the Lyman limit, the spectrum represents only noise with a flux around zero, similar to the quasar at z=1.618 in figure 5.16. At z=6, redshifted Lyman UV photons are leaving the visible range and reaching the infrared wavelengths at 7400Å as indicated.

It is interesting to see that for high z, an equivalent of the Lyman forest may be present redward of the redshifted Lyman alpha peak since the number of walls crossed is 660 for z=6 over a distance of about 86 billion l-y with consequent increase in absorption and reddening. Note that in the big bang model, at z=6, the distance traveled and stretched by photons is only about 31 billion l-y, which is 2.8 times less than in the multibang model, thus decreasing the projections of absorption and reddening in such a shorter distance model.

For example, quasar Q1351+6400 located at a redshift of z=0.088 has a flux of 3×10^{-13} erg cm^{-2}s^{-1}Å^{-1} at the Lyman alpha peak, and quasar SDSS 163033.90+401209.6 at z=6.05 has a flux of 3×10^{-17} erg cm^{-2}s^{-1}Å^{-1} at the Lyman alpha peak. This represents a difference of 4.0 orders of magnitude for a z distance 69 times farther away. Applying the inverse square law, we find that the signal from the far quasar should be 4,750 times lower, which is about 3.7 orders of magnitude lower, the small difference being attributable to reddening. But as mentioned, there are large variations of brightness between quasars with the same redshifts.

Figure 5.20 Quasar redshift versus Lyman alpha flux, compared with three inverse laws

The flux for the Lyman alpha peak from 156 quasar spectra [86] have been extracted and binned for averages in z intervals of 0.05. The result is displayed as a dotted line in figure 5.20 where the horizontal scale represents the redshifts from z=0.05 to z=1.55, and the vertical scale is the flux in erg cm^{-2} s^{-1} Å^{-1}. The Lyman alpha maximum flux is compared with three different inverse laws that have been associated with the decrease of brightness with distance: the dash line represents the static model and corresponds to simple inverse values expressed as $1/(1+x)$, where x is the redshift rank number in multiples of the basic interval z=0.05; the solid line corresponds to the multibang model without absorption and is represented by the inverse square law expressed as $1/(1+x)^2$; and the dash-dotted line represents the inverse cube values expressed as $1/(1+x)^3$. The big bang model is not represented but would correspond to the inverse fourth power law as $1/(1+x)^4$ and be drawn below the inverse cube curve way outside of the range of recorded fluxes.

The Lyman flux varies as the inverse cube of the distance for the first three values up to a distance of about 2–3 billion l-y, and it gradually crosses over and moves closer to a normal variation as the inverse square of the distance at about 9 billion l-y. The variation as $1/(1+x)^3$ for the first values corresponds to the 10.2 to 13.6 eV high Lyman absorption by neutral hydrogen up to redshift z=0.33. Then absorption decreases gradually to $1/(1+x)^2$ as the standard inverse square law over larger distances.

The flux is close to or slightly above the inverse square law in the right half of the curve up to the limit at 22 billion l-y. The latter is likely due to quasar selection effect or the anthropic effect since brighter quasars are more easily detected and preferentially selected, thus giving a higher flux. Some prior evaluations for the Tolman test have found inconclusive brightness variations between $1/(1+z)$ and $1/(1+z)^{3.4}$ from faraway sources.

6. Dynamics of Galaxies and Clusters of Galaxies

Galactic rotation velocities measured from Doppler shifts and distances to center of mass present flat curves away from the center and up to the edge of most large spiral galaxies. A similar problem is found in clusters of galaxies where apparent orbiting or infalling galaxies present excessive redshift velocities incompatible with dynamically stable systems. Cumulative masses calculated near the edge of these galaxies from flat rotation curves exceed what is expected from Kepler-Newton dynamics but can be accounted for if corrected gravitational forces are computed and applied to rotation velocities of galactic disks. Kepler-Newton laws apply only for idealized point-like masses or uniform-density spherical masses where force decreases as the inverse square law and where the geometric center and the center of mass coincide. In nature, all masses have dimensions, and none is perfect since field frictions are present everywhere [39].

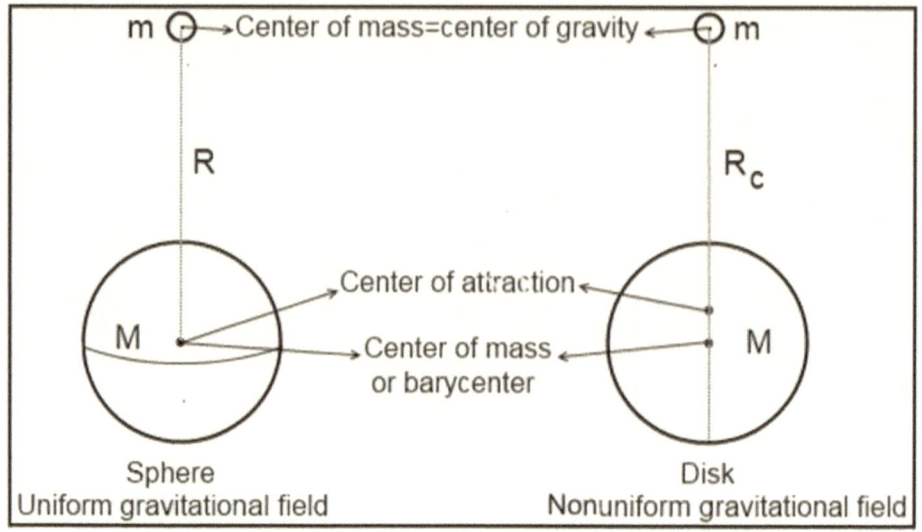

Figure 6.1 Comparison of gravitational attraction from a sphere and from a disk

A comparison of the gravitational attraction from a sphere and from a disk is presented in figure 6.1. On the left, a sphere with a large mass M generates a uniform gravitational field in all directions, so for a small body with point-like unit mass m, the center of mass and the center of gravity are located at the same point. Since m is negligible, the center of mass of the sphere and the barycenter of the system are located at the center of the large sphere, where the center of attraction is also found. The center of attraction is defined as the point toward which bodies tend by gravity. The term *center of attraction* is used for a clear distinction from the term *center of mass* that is generally related to the average location of the weight of a body or object.

But for a homogeneous disk made up of a large number of separate elements, as shown on the right, the gravitational field is not uniform in all directions although the center of mass of the disk and the barycenter of the system are located at the center of the disk. For a uniform disk, the center of attraction is located on the disk and along the line across the center of the two bodies. It may be located closer to orbiting mass m and yield a spiraling-in orbit or at the center of mass and produce a circular or elliptical orbit or on the other side of the center of mass and produce a bar. R_c represents the distance between the small orbiting mass m and the center of attraction located on the disk. The orbiting mass is rotating about the point at the center of attraction, while some separate bodies within the disk and closer to the center may simultaneously be orbiting about the center of mass.

The force is zero everywhere inside a perfectly uniform spherical shell. In this section, we show that for a uniform-density disk and except at the geometric mass center, the force is not zero inside a ring and larger than the inverse square law away from the edge of the disk. Doppler shift velocities and equivalent Keplerian velocities, combined with corrected force and related center of mass or center of attraction radiuses, will yield corrected masses corresponding to the volume density of physical entities such as galaxies and thus will be more compatible with other methods of mass evaluation like mass/luminosity distributions. No need to modify classical dynamics or to account for unknown dark matter [40].

The model will be expanded to a stack of disks and applied to four typical concrete examples, which are the large galaxy NGC 3198, with a disk radius of 30 kpc; the small galaxy NGC 4062, with a disk radius of 4 kpc; the barred galaxy NGC 4389, with a disk radius of 6 kpc; and the elliptical galaxy NGC 3379, with a disk radius of 12 kpc. Then the results will be extended to a more statistical number of 46 galaxies in order to extract some significant features, for example, about low- and high-surface brightness galaxies. A linear string of masses will also be discussed to understand the more elongated filamentary structures.

6.1 Disk Model and Equations

In the absence of an ideally large three-dimensional database to study the gravitational relationships between billions of masses with complex orbits in spiral galaxies, a simplified model comparing a galaxy to a disk or a stack of disks is presented.

The relationship between a unit test mass located at the edge of a disk and this same disk divided in a sequence of radiuses and a set of angles is discussed in terms of gravitational forces, velocities, masses, and ratios. The test mass will also be gradually moved across the disk and away from the edge of the disk to measure the importance of velocity anomalies and their effects on mass computations.

Equations related to the disk model are presented. Then graphs and results are discussed until the method is demonstrated. The few examples presented, including galaxy NGC 3198, will be computed with a stack of up to 30 disks to account for density variations across the galactic disk.

All gravity-related equations applied are from the Kepler-Newton classical physics and without modifications. Another set of equations related to geometry and trigonometry between unit mass and disk is used and developed here to measure total gravitational force from the total mass and cosine-corrected gravitational force from the truncated rings divided in point-like masses.

Finally, matrices are presented to compute any number of *m* rows (rings) by *n* columns (disks) in practical applications to galaxies. Other methods of summation may be used to obtain similar results.

6.1.1   Disk Model Presentation

In order to compute the total and the cosine-corrected gravitational force between a unit test mass and a disk, the following geometric model is used with appropriate symbols.

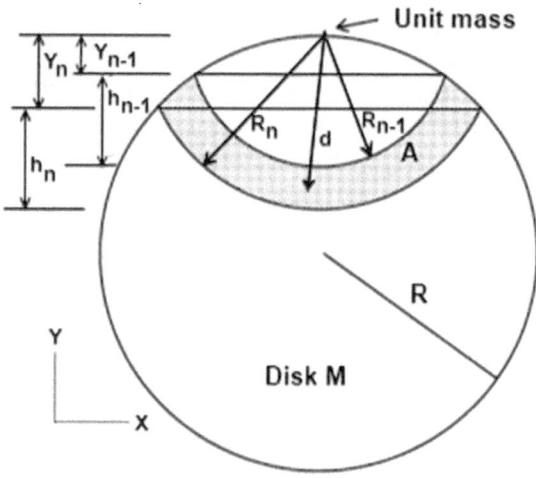

Figure 6.2 Disk *M* under study with segment divisions

Disk M is shown in figure 6.2 with a horizontal x-axis and a vertical y-axis. The unit mass on the edge of the disk is the center from which R_n variable radius values are generated to cover the whole disk with thin, truncated circular ring areas like *A* (in gray). Since the distance between the unit mass and any ring is constant, then the force between the unit mass and equal masses in the ring is constant, and the ring may be divided in equal point-like masses. Distance *d* in the middle of the ring is midway between any R_n and R_{n-1} values. Radius *R* is the fixed radius of disk *M* under study. The upper part of figure 6.2 is reproduced in figure 6.3 to show areas added and subtracted to obtain the truncated ring area *A*.

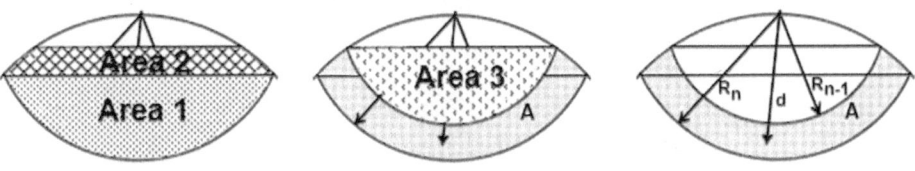

Figure 6.3 Truncated ring computations.

Area A is the truncated part of a flat ring in a disk and is equal to the segment with height h_n (area 1), plus the flat area between the two chords with height Y_n and Y_{n-1} (area 2), and minus the segment with height h_{n-1} (area 3). The idea is to use equidistant R_n values and associated masses to compute the gravitational force between the test mass of one unit and every ring to obtain total gravitational force values F_t for each ring and ΣF_t for the whole disk. At the same time, cosine-corrected gravitational force values F_c from every ring and ΣF_c for the disk will be computed.

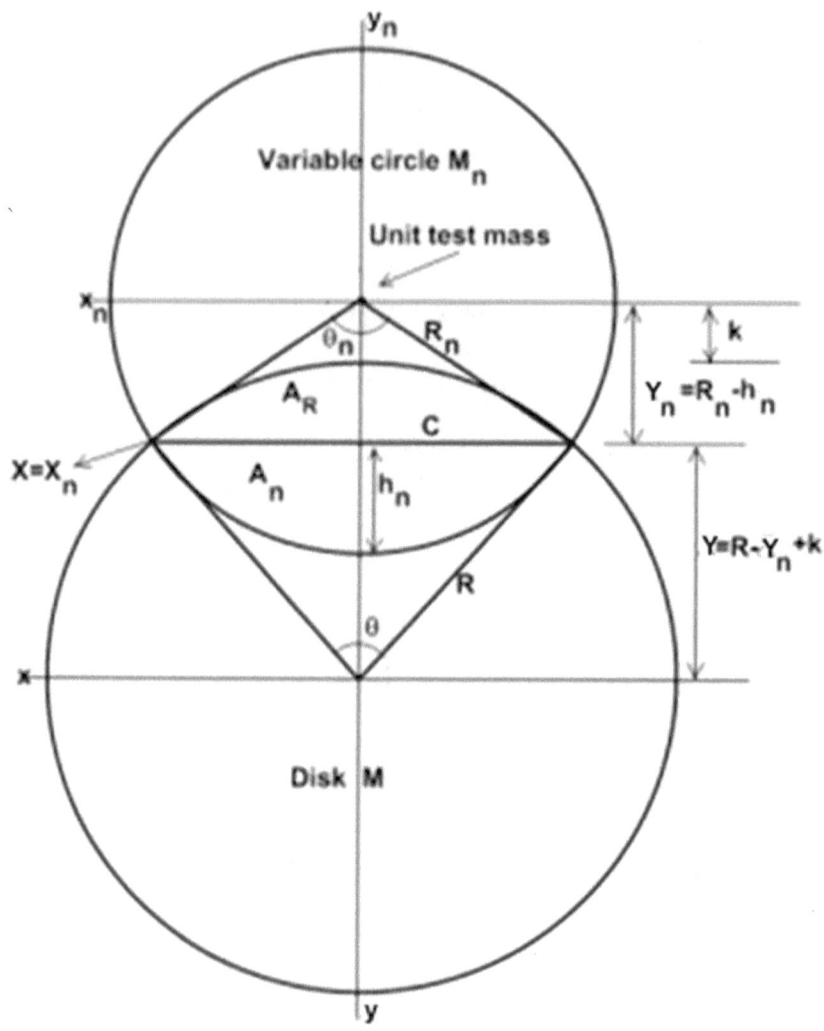

Figure 6.4 Disk M and variable circle M_n with associated symbols

Variable circle M_n is presented in figure 6.4 at a distance R_n from the unit test mass, with its horizontal x_n axis and its vertical y_n axis. Angle θ_n is formed between any two R_n radiuses running from the origin of circle M_n to the extremities of chord C. Likewise, angle θ is between two R radiuses and runs from the origin of disk M to the end of chord C. Area A_n with height h_n is a part of circle M_n limited by chord C. Area A_R is a part of disk M also limited by chord C.

The height of *k* represents the distance between the unit test mass and the edge of disk *M*. When k=0, the test mass is located right on the edge of disk *M* as shown in figure 6.2. When *k* is positive, as in figure 6.4, the test mass is moving away from the edge; and with *k* negative, the test mass is moving across the disk toward the center.

6.1.2   Circle and Disk Equations

From the graph of figure 6.4, we have

$$Y=R-Y_n+k \qquad (6.1)$$

and

$$Y_n=R_n-h_n. \qquad (6.2).$$

The extremities of chord *C* are located at the junction of disk *M* with circle M_n, where $X=X_n$.

From the equation of the circle, $X^2+Y^2=R^2$ and $X_n^2+Y_n^2=R_n^2$, and from equations 6.1 and 6.2, we have

$$R_n^2-Y_n^2=R^2-Y^2 \qquad (6.3).$$

Replacing *Y*, we find

$$Y_n=(R_n^2+2Rk+k^2)/(2R+2k) \qquad (6.4).$$

And replacing Y_n gives

$$h_n=R_n-((R_n^2+2Rk+k^2)/(2R+2k)) \qquad (6.5).$$

Area *A* of the truncated ring is explained in figure 6.2 and combined to figure 6.4 to obtain

$$A=A_n-A_{n-1}+(A_R-A_{R-1}) \qquad (6.6),$$

where A_{n-1} and A_{R-1} are the preceding values of A_n and A_R, respectively, and correspond to radius R_{n-1}.

Available circle formulas may be combined to the above geometry and to trigonometry to find angles, chord lengths, sector, and segment areas as follows:

Sector area for circle M_n is found from

$$\text{sector area}=R_n^2(\theta_n/2) \qquad (6.7)$$

and for disk *M* from

$$\text{sector area}=R^2(\theta/2) \qquad (6.8),$$

where the angle for M_n is

$$\theta_n = 2 \arccos(Y_n/R_n) \tag{6.9}$$

and for disk M

$$\theta = 2 \arccos(Y/R) \tag{6.10}.$$

The area of the two triangles above chord C in figure 6.4 is

$$M_n \text{ triangles} = Y_n(C/2) \tag{6.11}$$

and for the two triangles below chord C

$$M \text{ triangles} = Y(C/2) \tag{6.12},$$

where $C/2$ is defined as

$$C/2 = R_n \sin(\theta_n/2) \tag{6.13}$$

or

$$C/2 = (R_n^2 - Y_n^2)^{1/2} \tag{6.14}.$$

Then area A_n in circle M_n is equal to the sector area, equation 6.7, minus the two triangle areas, equation 6.11, combined with equation 6.14:

$$A_n = R_n^2(\theta_n/2) - (Y_n(R_n^2 - Y_n^2)^{1/2}) \tag{6.15}.$$

And area A_R in disk M is similarly expressed as

$$A_R = R^2(\theta/2) - (Y(R_n^2 - Y_n^2)^{1/2}) \tag{6.16}.$$

Equations 6.15 and 6.16 and equivalent equations for A_{n-1} and A_{r-1} will be introduced in equation 6.6 to determine area A of the truncated ring for each R_n radius value across the entire disk:

$$A = [R_n^2(\theta_n/2) - (Y_n(R_n^2 - Y_n^2)^{1/2})] - [R_{n-1}^2(\theta_{n-1}/2) - (Y_{n-1}(R_{n-1}^2 - Y_{n-1}^2)^{1/2})]$$
$$+ [(R^2(\theta/2) - (Y(R_n^2 - Y_n^2)^{1/2})) - (R^2(\theta_{-1}/2) - (Y_{-1}(R_{n-1}^2 - Y_{n-1}^2)^{1/2}))] \tag{6.17},$$

where subscript -1 indicates the previous value.

The sum of all the ring areas, ΣA, is the surface of the disk. The volume and mass of the disk is the same value as the surface since a thickness of one unit is used as well as a density of one unit to simplify the model. Furthermore, the gravitational constant, G, may be left aside since it does not affect ratios in the model, and it can be reintroduced at will in concrete examples. The purpose is to find ratios to evaluate velocity and mass relationships between a test mass of one unit and the sum of the rings of increasing radiuses across a whole disk and then transpose the results to galactic scale. In the model, the ratio between the radius and the thickness of the disk varies from one to 150 because most galaxies seem to have a ratio radius/thickness less than 150.

Figure 6.5 Mass computed from volume and density at the edge of disks increasing in size and away from the edge of the largest disk

The curve in figure 6.5 shows the computed increasing mass values at the edge of disks increasing in size from a radius R=0 on the left to R=150 distance units on the right, where it flattens out away from the edge of the largest disk over a distance of 50 units for a total distance of two hundred units. Values, starting at zero, are computed every five distance units for a total of 41 computations across the horizontal scale. Mass is additive and increases with disk radiuses until it flattens out to a maximum at the edge of the largest disk. It is expressed in units on the vertical scale. This calibrated mass curve is calculated with the volume-density equation

$$M = \pi R^2 t_d \rho \qquad (6.18),$$

where R=0 to 150 is disk radius, $t_d=1$ is the thickness of the disk, and $\rho=1$ is the density. So in this model, masses are simply equal to the area of the disks.

Two kinds of gravitational forces, two different velocities, and two radial distances associated with the mass distribution in figure 6.5 will be studied, and the same mass distribution as in figure 6.5 will be found with the application of cosine-corrected force and radius to Doppler shift velocities to compute galactic masses.

6.1.3 Cosine-Corrected Gravitational Force

Equation (6.17) is applied to every truncated ring in the disk to obtain separate area, volume, and mass for the rings. Total gravitational force for each ring is then computed from Newton's equation,

$$F_t = GMm/d^2 \qquad (6.19),$$

where F_t is the total gravitational force per ring, G is the gravitational constant, M is the mass of the disk per ring, m is the test mass equal to one, and d is

$$d = (R_n + R_{n-1})/2 \qquad (6.20).$$

Errors from averaging d in the middle of R_n intervals are proportionately lower when a larger number of R_n values are used.

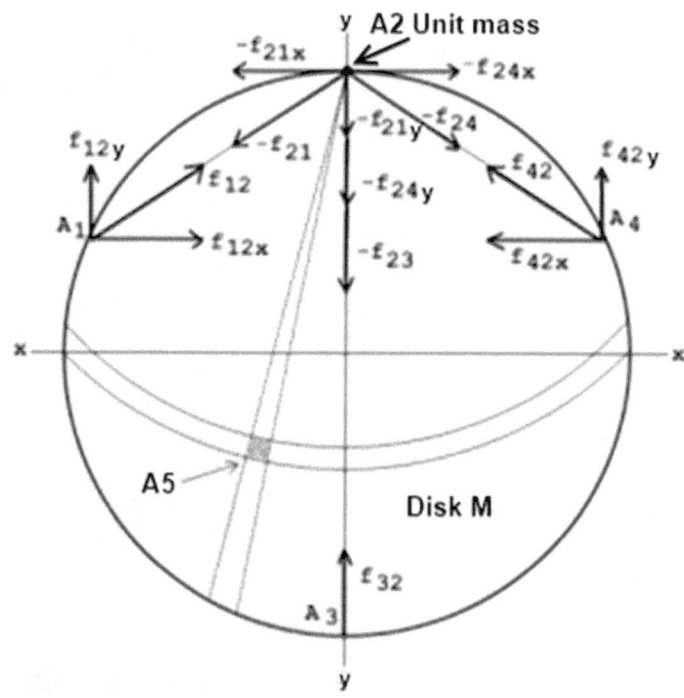

Figure 6.6 Cosine angles across the rings and vectors

One truncated ring formed by the interval between two sequential R_n values is reproduced in figure 6.6. It is intersected at $A5$ by a small portion of θ_n angle. Data sampling is accomplished by dividing the whole disk in small point-like masses. The size of data samples for a fixed angle increases away from the unit mass with distance, but this is compensated since force decreases as the inverse square law. Figure 6.6 also shows a graphical example of the vector aspect of gravitational force. When unit mass is located at point $A2$, it attracts each of the three point-like unit masses located at A_1, A_3, and A_4, as shown with the vectors and their x and y components.

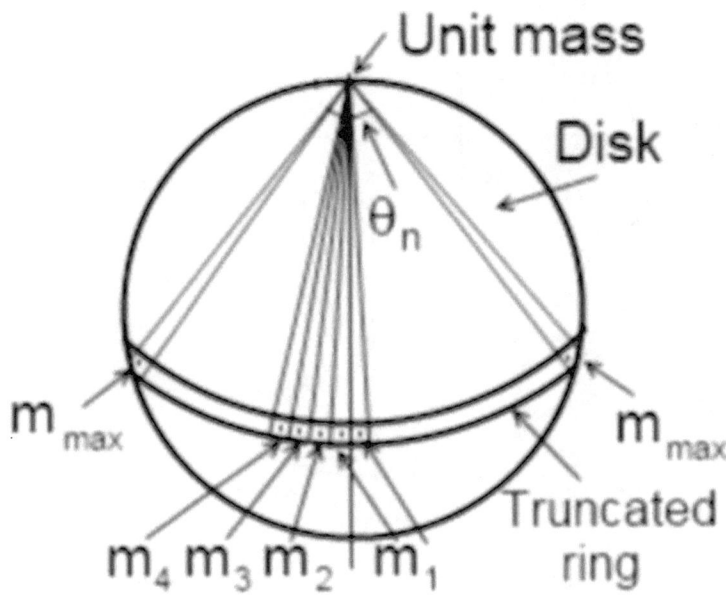

Figure 6.7 Cosine summation for one truncated ring

As shown in figure 6.7, all truncated rings are divided in small portions, m_1, m_2,..., up to m_{max}, by a large number of equally spaced angle subdivisions that may be selected arbitrarily in proportion with the number of R_n subdivisions. Angle θ_n has a maximum of 180 degrees or π radians, and its subdivisions are symmetric on each side of the y-axis. Force vector values will be found from the cosine of each subangle. Since mass is equally subdivided within every ring, it is possible to sum all the cosine values from subangles crossing a ring with the following equation,

$$\text{Cosine sum} = \sum_{m=1}^{m_{max}} \cos[\theta_n(((2m)-1)/(4m_{max}))] \quad (6.21),$$

and obtain corrected gravitational force, F_c, for that ring if we multiply the cosine sum by the point-like mass within one subangle. Then the sum of F_c force vector values for all rings, ΣF_c, will give the cosine-corrected force for the whole disk.

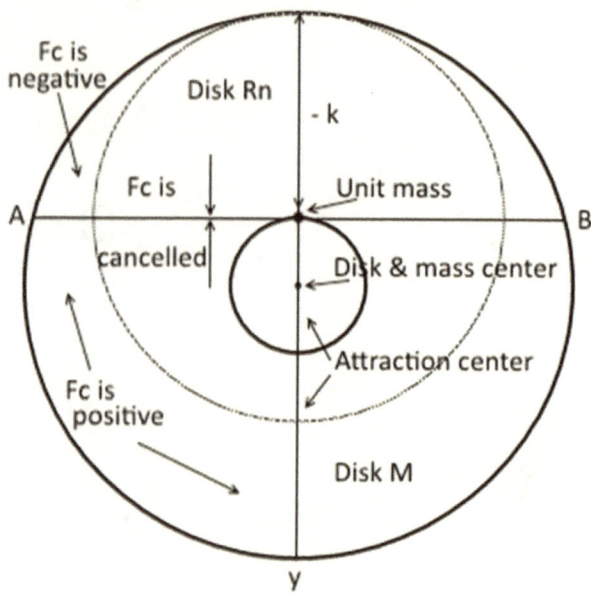

Figure 6.8 Corrected force F_c when unit mass is moved across the disk

When the unit mass is displaced across the disk along the y-axis, the configuration to compute the corrected force F_c is gradually different and represented in figure 6.8. The unit test mass is located at a distance k from the top edge of disk M for which disk and mass center are shown. Disk R_n is the largest disk around the unit mass that is completely enclosed in disk M. Force vectors between the unit mass and all point masses in disk R_n cancel out on opposite sides of line AB, so the net corrected force F_c is calculated between unit mass and all point masses located in the crescent of disk M, where F_c will be negative above line AB and positive below that line. Since the positive portion of the crescent is the most extended, the net result should be a positive force F_c.

The center of attraction of the latter force will be located across the mass center of disk M when the unit mass is located close to the center of the disk, and then the orbiting mass will spiral out and stimulate the formation of a bar. But when the center of attraction coincides with the center of disk M, we expect the unit mass to orbit in a circle or in an ellipse; and when the center of attraction is located between the unit mass and the center of mass of disk *M*, we expect the unit mass to spiral in toward the center.

6.1.4 Graphs and Results

Most of the graphs and results will be for disks in the range R_n=0 to 150 units, which includes radius/thickness ratios similar to those for a majority of regular spiral galaxies. But an additional distance ranging from k=5 to 50 units will be presented to show what occurs away from the edge of the disk when no additional mass is present. Negative *k* values will also be explored across the disks. A total distance of 200 units in steps of five units will yield 41 values in the following theoretical model.

Figure 6.9 Comparison between cosine-corrected gravitational force F_c across 30 disks. Also, F_c across and away from the largest disk with a radius of 150 units (triangles) and Newton's gravitational force F_n (squares)

The horizontal scale is the same as in figure 6.5, and the vertical scale shows gravitational force in relative units. The curve with diamonds represents net force vector or cosine-corrected gravitational force, ΣF_c, between unit mass at the edge of a disk and the mass within the disk for every selected disk with radiuses increasing in steps of five units. Force increases up to the edge of the last disk at a distance of 150 units. Averaging and edge error is larger on the first values but decreases rapidly as the cylinder flattens to a disk. This curve is computed, as shown above, for every point-like mass of the disk with Newton's equation 6.19 and cosine corrections. For example, a disk with R=150 could have R_n values from 0 to 300 and θ_n divisions from 0 to 100 for a total number of 30,000 computation points on the disk. Note that Newton's gravitational equation, which applies directly only to point-like masses or homogeneous spheres, is valid in this case since the disk is divided in point-like masses.

The curve with triangles represents the cosine-corrected force, ΣF_c between the whole disk of 150 units and the unit mass moved across the disk for every selected radiuses increasing in steps of five units. Force increases from zero at the left and up to a maximum at the edge of the disk at a distance of 150 units, where it is the same as the last value of the curve with diamonds. Then it decreases sharply away from the edge. This curve shows values lower than force F_n curve (squares) near disk center and crosses it before reaching a maximum.

The curve with squares, F_n, represents the flat curve from equivalent center of mass force with a constant force value π up to the edge of the last disk at 150 units and then declining steadily away from the last disk. It is computed with Newton's equation 6.19, where disk mass increases as the square of the radius and force decreases as the square of the radius, thus yielding a constant value. But it is based on the false hypothesis that the mass of the disk is homogeneous like that of a sphere in 3-D. Contrary to the sphere, where Newton's equation can be solved with success, assuming that the whole mass of the sphere is at its center, this is not the case with a flat disk. Note that gravitational forces F_c and F_n are both 0, as shown on the graph, at the center of any homogeneous disk because of the symmetric cancellation of all the vectors.

Figure 6.10 Gravitational force F_c across 30 different disks with radiuses from 0 to 150 units and away up to 200 units in steps of five units

Horizontal and vertical scales are the same as in figure 6.9. Last curve to the right in figure 6.10 (diamonds) is the same as the curve with triangles on the previous figure. Maxima of these 30 curves correspond exactly to the curve with diamonds from disk edges in figure 6.9. Each curve starts with zero at the center of a disk, reaches a maximum, and decreases toward zero away to the right. These *n* disks, with a density of one, may be combined with their *m* rings to form matrices from which forces, velocities, densities, and masses will be computed for real galaxies.

Figure 6.11 Ratios: center of mass radius over center-of-attraction radius, R/R_c, at the edge of 30 disks (diamonds), across the disk with radius R=150 (squares), and Newtonian (triangles)

Horizontal scale is the same as in figure 6.10, and vertical scale represents ratios of center of mass radius over center-of-attraction radius, R/R_c.

The curve with diamonds ranges, in figure 6.11, from 1.0 at the center of the disk on the left to a maximum of about 2.2 at a distance of R=150 on the right. The curve with squares also starts from 1.0 and varies, across the largest disk with R=150 units, from a minimum of about 0.2 to a common maximum near 2.2, and then it swings down toward 1.0 away from the edge of the disk. Newtonian ratios (triangles) for the 30 disks stay flat at 1.0 all the way.

A value of one indicates that center of mass and center of attraction coincide, and a value of 2.0 indicates that the center of attraction is displaced from the center of mass to a location halfway between the test mass and the center of the disk. A lower value than one indicates that the center of attraction is displaced to the opposite side of the center of mass. The latter is directly related to the control of bar formation in spiral galaxies. Such displacements have consequences on velocity and mass computations.

Center of attraction at a distance R_c is computed with Newton's gravity equation 6.19, where both mass of the disk and cosine-corrected force of attraction, ΣF_c (top curve in figure 6.9), are known,

$$R_c = (MG/\Sigma F_c)^{1/2} \quad (6.22),$$

where R_c is the radial distance from the test mass. Constant G is neglected on the curves.

Or for a single ring: $\quad R_c = (MG/F_c)^{1/2} \quad (6.23).$

Figure 6.12 Top curve is Doppler velocity, V_c, and Keplerian velocity, V_K, is the bottom one

Horizontal scale is as above, and vertical scale represents velocity units. Each of the two velocities, V_c and V_K, in figure 6.12 increases from the center at left and across the 30 disk edges, up to the largest at R=150 on the right, and then drops down toward converging values away from the last disk.

The top curve is Doppler equivalent or corrected velocity, V_c, and it is the largest, about one and a half times the Keplerian velocity in this example, and would correspond in real life to best fit Doppler shifts velocity evaluation across a galactic disk. V_c is also the corrected velocity calculated from corrected force, center of attraction radius and/or corrected mass. Thus, Doppler and corrected velocities are the same values and either name or both can be used. Velocities from Doppler shifts are currently used by astrophysicists to compute galactic masses and yield large errors in mass evaluations when combined with center of mass radius. These excessive mass errors are at the origin of the mysterious and elusive dark matter concept invented in the big bang model but never proven.

But, this is the velocity at which the orbiting mass is moving around the center of gravitational attraction located at R_c. The latter is located either closer to, at, or farther away from the orbiting mass than the center of mass. This velocity may be computed with the circular velocity equation

$$V_c=(GM/R_c)^{1/2} \qquad (6.24),$$

where G is the gravitational constant which is neglected, M is the mass inside the volume of the disk, and R_c is the corrected distance between the orbiting mass and the center of attraction, which varies according to the ratios in figure 6.11. So, this is the velocity that will yield the corrected or real mass when combined with the corrected distance.

Finally, the Keplerian velocity, V_K, will also give the real mass inside a disk since it is based on the real total mass and total radius of the disk. The circular velocity equation 6.24 is also used to compute this velocity,

$$V_K=(GM/R)^{1/2} \qquad (6.25),$$

where G and M are the same as for the previous velocity, and R is the distance between the orbiting mass and the center of mass, which is also the geometrical center of the disk. The test mass is orbiting at this velocity about the center of mass. V_K is the velocity needed in an equivalent spherical Keplerian dynamic system to find correct mass values.

Figure 6.13 Comparison of corrected velocity, V_c (squares); and Keplerian velocity, V_K (triangles), for single disk of 150 units

Horizontal scale is as above, and vertical scale represents velocity units. Each of the two velocities, V_c, and V_K, in figure 6.13 increases from the center of this constant density disk at left to the edge at R=150 on the right and then drops down toward converging values away from the edge. The maximum of these curves are the same values as for the disk with R=150 in figure 6.12.

Squares represent velocity, V_c, that corresponds in real life to best fit Doppler shifts' velocity evaluations across a homogeneous galactic disk. Since this curve crosses the other curve, it means that velocities, when combined with center of mass radiuses, yield undervalued masses near the center of the disk and overestimated ones toward the edge. But the most important observation is that, for a uniform disk, velocity increases gradually from the center of the disk toward the edge where the increase is very sharp. So no wonder we find flat or increasing Doppler shift velocities toward the edge of a galactic disk.

This velocity, V_c, will yield the corrected mass when combined with the corrected distance.

Finally, the Keplerian velocity V_K (triangles) will also give the real corrected mass inside a disk since it is based on the real total mass and total radius of the disk. V_K is larger than V_c near the disk's center but crosses over and remains lower near the maximum and away from the edge. Note that, for a uniform disk, velocity V_K also increases gradually from the center of the disk and all the way to the edge.

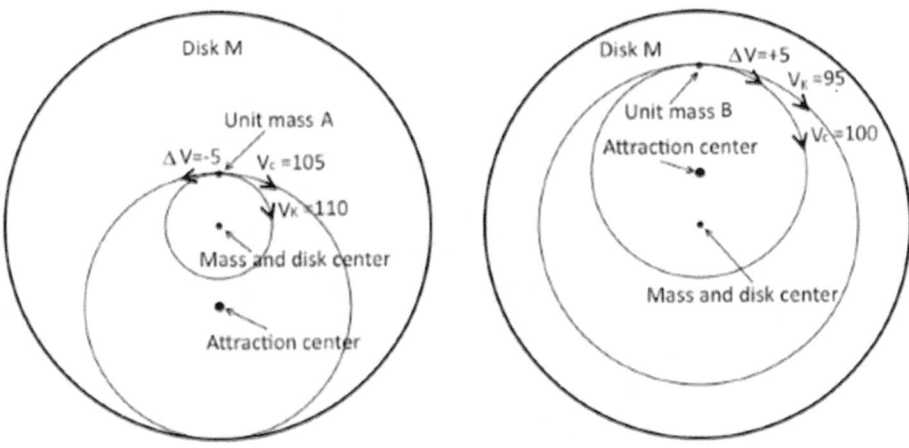

Figure 6.14 Example of an orbiting mass A spiraling out on the left and mass B spiraling in on the right

The $\Delta V = V_c - V_K$ velocity vector difference between V_c and V_K is critical for the orbiting material around the disk. Masses are really orbiting around the center of attraction with the corrected or Doppler velocity V_c but they are simultaneously carried by the overall galactic rotation around the center of mass of the disk with a velocity V_K so that the ΔV velocity vector represents the excess velocity around the center of attraction. Note that acceleration is directed toward both center of attraction and center of mass. In figure 6.14, we show what happens to the left and to the right of the crossover in figure 6.13 with numerical examples attributed to V_c and V_K. To the right of the crossover in figure 6.13 and to the right in the example of figure 6.14, ΔV is positive since V_c is larger than V_K which indicates that orbiting masses will spiral in, as shown in the example, at a velocity $\Delta V = +5$ toward the center of the disk and directly contribute to form the related and impressive spiral arms within the galactic disk.

When the vector ΔV is equal to zero at a crossover, then orbiting masses will form either an ellipse or a circle around the center of the disk, but no spiral since the center of attraction and the center of mass coincide.

To the left of the crossover, ΔV is negative since V_c is smaller than V_K which means that orbiting masses around the attraction center are lagging and will spiral out in the opposite direction at a velocity, for example, of $\Delta V = -5$ away from the center of the disk and build up an elongated bulge followed by a bar which is generally much less impressive than spiral arms covering a larger surface and volume.

Figure 6.15 Inverted colors of low surface brightness galaxy NGC 4921 (Credit: NASA/ESA/HST)

An inverted black-and-white picture from the barred spiral galaxy NGC 4921 located at an estimated distance of 310 M l-y in the Coma cluster is presented in figure 6.15. The original picture of this low surface brightness and low star-forming galaxy was taken by the Hubble Space Telescope for NASA/ESA. The galaxy is leaving a trail of frayed dust and gas, as observed below the arrow indicating the ring of dust clouds, since it is moving at high speed toward its own center of attraction in the hot ionized gas of the Coma cluster. The galaxy has left the cocooning wall where it was born and has grown from the accretion of new star-forming neutral or molecular hydrogen. Both the gravitational attraction of the Coma cluster and the pressure of the residual kinetic energy of its expanding bubble have contributed to its cosmic journey.

NGC 4921 is a good example showing the direction of masses orbiting a galaxy as related to velocities V_c and V_K. As mentioned in figure 6.13 and 6.14, when ΔV is positive, then orbiting masses spiral in as shown in figure 6.15 and form the spiraling arms of the galaxy. At the crossover, ΔV is equal to zero, so masses are forming a circle or an ellipse around the center with the accumulation of gas and dust star forming material as in a traffic jam. When ΔV is negative, then orbiting masses spiral out and form an elongated bulge or a bar as observed.

Figure 6.16 Corrected mass M (M_c or M_K, squares); apparent mass (triangles) at the edge of 30 disks; and apparent mass (diamonds) across largest disk

The horizontal scales of figure 6.16 is the same as in figure 6.5, but the vertical scale has a larger range than that in figure 6.5, where the calibrated mass was computed.

Apparent mass M_a (triangles) at the edge of 30 different disks increases much faster than corrected mass M. It is computed with the circular velocity equation 6.24,

$$M_a = V_c^2 R/G \qquad (6.26),$$

where R is the radial distance, and V_c is the velocity measured from Doppler shifts in real life. Apparent mass values are 2.16 times larger than corrected mass at a radius of 150. This implies important errors of interpretation on the mass of galaxies such as the introduction of fictitious dark matter when apparent mass is used instead of corrected mass.

The maximum value of the latter curve, and away to 200 units, is the same as on the curve with diamonds, which represents apparent mass across the larger disk with R=150 and up to 50 units away from the edge. This curve crosses the curve of the corrected mass M, indicating that apparent mass would yield a mass deficit toward the center of the disk and a large mass excess toward the edge of the same disk. It shows a sharp break to lower values at the edge of the disk but remains larger than mass M while converging. Obviously, apparent mass is not real since it shows an impossible mass decrease away from the edge of the disk. Mass is positive and additive within a sphere or a disk and can only increase away from the center of a body or set of bodies. Negative mass was never recorded.

The curve with squares increases from 0 to 150 units and then is flat away from the edge of the last disk. It represents the cumulative calculated mass M, which is equal to corrected mass, M_c, or Keplerian mass, M_K, from velocity curves in figure 6.12. The circular velocity equation 6.24 may be used to compute this curve with $M_c = V_c^2 R_c/G$, where M_c is the corrected mass, V_c is the Doppler or cosine-corrected velocity, and R_c is the corrected distance, or with equation 6.25, $M_K = V_K^2 R/G$, where M_K is the Keplerian mass, V_K the Keplerian velocity, and R the radial distance. Note that symbol M will be used in both cases since $M = M_c = M_K$. This mass curve has identical values as the calibrated mass from the original disk model in figure 6.5 and is reproduced below as black bars in figure 6.17.

Figure 6.17 Mass at the edge of 30 separate disks and away as physically calibrated and after model corrections

The scales of figure 6.17 are identical to figure 6.5, but the display is made with adjacent black and gray bars. The black bars represent the calibrated mass to the edge of the disks and away as calculated with equation 6.18, and the gray bars represent the mass to the edge of the disks and away as calculated from the model with Doppler equivalent or corrected V_c or Keplerian velocities V_K. These black and gray bars are identical for each disk, thus confirming this method of computation based on truncated rings and cosine corrections model, whereby corrected or Keplerian mass values correspond to the real mass of a disk. Therefore, apparent mass values, as usually computed by astrophysicists from Doppler shift velocities, combined with center of mass radiuses, are erroneous and must be rejected with the related dark matter concept.

In a nutshell, the loops from mass M at the start to mass M at the end of this theoretical example may be summarized as follows in figure 6.18:

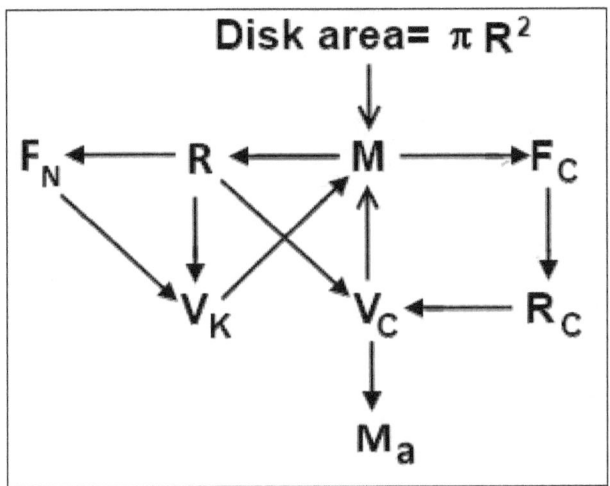

Figure 6.18 Summary of the theoretical model

From the radius of a disk, mass is found to compute the corrected force from which the corrected radial distance and Doppler or corrected velocity will be calculated and yield the original mass value or else will yield the apparent mass when combined with center of mass radius. The original mass, combined with R, also gives the Newtonian force and then the Keplerian velocity to find again the original mass. This completes the proof on the calibrated mass for the 30 selected disks with radius R=0 to 150 units and the single selected disk with radius R=150 units and k=5 to 50 units away from the edge of the disk, computed in 41 steps of five units, all with a thickness of one unit and a density of one.

These results may be generalized and applied to any flat disk. They may also be combined with many disk layers of various sizes and/or the presence of a central bar, core, bulge, and more or less curved spiral arms in a stack of disks to simulate all kinds of galaxies.

6.1.5 Mass, Velocity, and Distance Ratios

Combining mass, velocity, and distance values, it is possible to establish a few useful ratios applicable to our disk models as well as to all galactic disk computations.

Figure 6.19 Mass ratios, M_a/M, or distance ratios, R/R_c, at the top; and velocity ratios, V_c/V_k, at the bottom

Horizontal scale is the same as on figure 6.16. Vertical scale shows ratios. The two ratio curves shown in figure 6.19 are based on disk models with density 1 and radiuses between R=0 and 150 units computed in steps of five units and with 50 units away from the edge of the largest disk.

The bottom curve represents Doppler or corrected-to-Keplerian-velocity ratios V_c/V_k at the edge of 30 separate disks. It reaches a maximum near a value of 1.47 at the edge of the largest disk and then decreases.

The top curve with a maximum value of 2.16 corresponds to two identical curves showing the same ratios for apparent-to-corrected or Keplerian mass M_a/M and also center-of-mass to center-of-attraction distance R/R_c already presented in figure 6.11.

The circular velocity equations 6.25 and 6.26 give $M=V_K^2R/G$ and $M_a=V_c^2R/G$, from which apparent-mass-to-corrected-mass ratios M_a/M are found equal to the square of corrected-to-Keplerian velocity ratios V_c^2/V_K^2:

$$M_a/M = V_c^2/V_K^2 \qquad (6.27)$$

We take Newton's equation, $V_c^2 = GM/R_c$, and equation 6.22, where $R_c = (GM/F_c)^{1/2}$, and we combine with the circular velocity equations 6.25 and 6.26 as above to yield a very useful result: the square of Doppler or corrected-to-Keplerian velocity ratios are also equal to the ratio between the center of mass and the center of attraction, R/R_c,

$$V_c^2/V_K^2 = R/R_c \qquad (6.28)$$

from which center of attraction radius, R_c, can be determined when Doppler shift equivalent velocity, V_c; Keplerian velocity, V_K; and radius R are measured values from, for example, galactic studies and descriptions.

Then the centripetal force equation 6.23 and $V_c = (F_c R_c/m)^{1/2}$ yield the centripetal acceleration per mass unit:

$$a = V_c^2/R_c \qquad (6.29)$$

But for the equivalent spherical model with a Newtonian force, F_n, and a Keplerian velocity, V_K, we would find the Newtonian acceleration, $a_n = V_k^2/R$.

Also, combining Newton's equations for forces F and F_n, we find the following useful result: the square of center-of-mass over center-of-attraction ratios R^2/R_c^2 is equal to the ratio corrected-force-to-Newtonian-force,

$$R^2/R_c^2 = F_c/F_N. \qquad (6.30)$$

6.2 Application to Four Typical Galaxies

In order to apply the results of the theoretical model to real-life galaxies, a more appropriate model has been developed to imitate the gradually increasing mass or density of a galactic disk from the edge to the center. Our method is using a stack of up to 30 disks to make it possible to calculate the mass of most spiral and elliptical galaxies with a good precision.

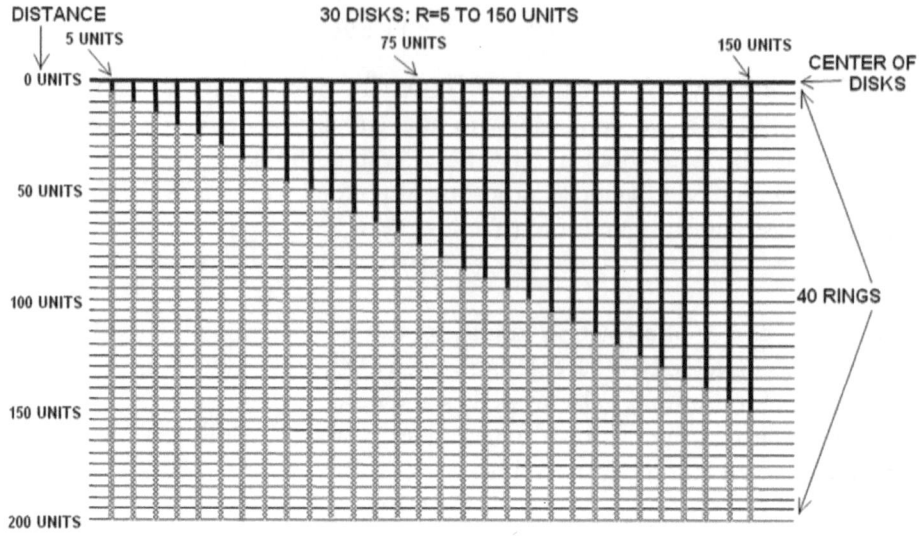

Figure 6.20 Matrix setup to compute galactic forces, velocities, densities, and masses using 30 disks with R=0 to 150 units in steps of five units

The basic m by n matrix setup used to compute forces, densities, masses, and velocities for spiral galaxies is shown in figure 6.20, with n=30 disks represented by black vertical lines with radiuses varying from 0 to 150 units in steps of five units. Three of these disks are specifically indicated as 5, 75, and 150 units. Disks correspond to the presence of masses starting at the center (top horizontal line) and extending down to their respective edges, maximum 150 distance units. Gravitational forces are present away from these disks and represented by gray line extensions away from the edges up to a maximum of 200 units in steps of five units. Horizontal lines correspond to m=40 rings, including 30 rings for the largest disk up to 150 units and 10 rings for its extension up to 200 distance units. When a value of zero is included at the center, a total of 41 rows is involved.

Three basic matrices, A, B, and C, are combined to compute the stack of the total force per ring, total mass, and total density for the disks. Matrix A (41 rows by 30 columns) contains gravitational force coefficients F_c for each ring element of each disk with a density of one and their extensions away from the edges. Matrix B (41 rows by 30 columns) contains density coefficients, ρ, found iteratively by matching velocities from Doppler shifts with computed corrected velocities using equation, $V_c=(\Sigma F_c R_c/m)^{1/2}$. Matrix C (31 rows by 30 columns) contains basic mass coefficients for each ring element of each disk.

Newton's equation says that force, F, is proportional to mass, M, implying that any increase of mass by a density factor will increase the value of force, F, in the same proportion, so multiplying each element of matrix A by the corresponding element of matrix B yields matrix D, which represents total force for each element. These elements are stacked horizontally to get the sum of total forces ΣF_c. Also, multiplying each element of matrix B (density) by the corresponding element of matrix C (mass) will yield matrix E, which represents total mass for each ring element of each disk. Again, these elements are stacked horizontally to get the sum of the total mass in each galactic ring. Finally, the mass of each ring is summed to obtain corrected or real galactic mass from which corrected-Doppler and Keplerian velocities are computed simultaneously in the same iteration process on a spreadsheet.

6.2.1 Application to a Large Galaxy: NGC 3198

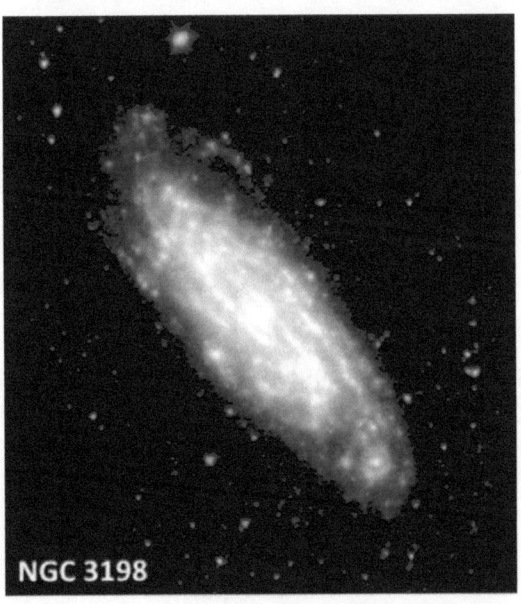

Figure 6.21 Black and white image of NGC 3198 as seen by the Spitzer space telescope (Credit to: NASA/JPL Caltech)

Galaxy NGC 3198 is a high-surface-brightness galaxy located in the Ursa Major constellation about 14.4 Mpc away, with a radius of about 29.7 kpc and classified as a spiral with a small core, an Sc galaxy. A black-and-white version of the galaxy is shown in figure 6.21 in which NGC 3198 shows bright blobs or spots corresponding to enhanced dusty regions where stars and star clusters are formed. The spiraling arms are clearly visible, and the bulge may suggest the beginning of a bar. The original color photo was taken by the infrared Spitzer space telescope for NASA/JPL Caltech. We assume in all theoretical and applied galactic models that no matter is present beyond the edge of the disk although we know that some uncharted gas is generally present and poorly evaluated.

Figure 6.22 Newton force and corrected force for NGC 3198

Galactic radius for NGC 3198 runs from left to right on the horizontal scale, from 0 to 30 kpc, with an extension of 10 kpc away from the edge of the disk. Note that horizontal distance scale is converted from the previous theoretical model with a factor of five units for 1 kpc. The vertical scale represents force in units in figure 6.22.

The distribution of Newton force, F_n, is shown with the square symbols: it is computed with Newton's universal law of gravitation and assumes that the mass across the disk is concentrated at the center and thus decreases with the square of the radius. On the other hand, the diamond symbols represent the corrected force, F_c, which is computed from the summation of the force distribution in a stack of 30 disks. Except near the center where an important mass must be present in the bulge and must give some similitude to the curve from Newton's force, it generally shows a larger and gradual decrease of force F_c across the galaxy corresponding to a gradual decrease of mass across the disk. We find $F_n = F_c$ at the center and at the crossover, $F_n > F_c$ between the center and the crossover, and $F_n < F_c$ between the crossover and the edge of the disk and even away from the disk.

Figure 6.23 NGC 3198 corrected velocity, V_c (squares); and Keplerian velocity, V_K (triangles) from a stack of 30 disks

The horizontal scale in figure 6.23 is the same as on the previous figure, and the vertical scale shows velocity in m/s.

The curve with squares from 0 to 30 kpc represents rotation velocities as extracted from Doppler shift measurements. An average is presented here from one of the most published velocity curve (Begeman, 1989) and plotted against distance from the center in kpc. Note the typical flat Doppler shift velocity curve associated with large galactic disk. The last 10 values are computed away from the edge of the galaxy from equation 6.23.

It corresponds to corrected-Doppler velocity V_c. It may be computed from the circular velocity equation 6.24 with $V_c=(GM/R_c)^{1/2}$, where M is corrected mass and R_c is center-of-attraction radius. These velocity values, as for the theoretical model, yield corrected mass for a disk when combined to center-of-attraction values. These are the velocities at which bodies are orbiting around their own center of attraction across the galactic disk.

The Keplerian velocity curve (triangles) represents the equivalent Keplerian spherical dynamics for NGC 3198 as if the center of mass were the center of attraction. It is calculated from cumulative matrix E corrected mass values. The last 10 Keplerian velocity values are also calculated away from the edge. When reduced to a spherical model, masses seem to be orbiting with velocity V_K around the center of mass although they really orbit according to V_c around R_c, the ΔV vector difference being proportional to the velocity down the spiral curve toward the center. This velocity curve is generally lower than the other one for V_c except for the first three values near the center of the galaxy, which implies that Doppler shifts velocities, when combined with center of mass radiuses, are too small near the center and too large toward the edge to obtain a correct mass evaluation. Note that this curve is gradually decreasing to the right after a maximum and is not a flat curve like that from Doppler or corrected velocities, thus indicating the absence of mass excess or dark matter toward the edge of the disk. But large clouds of undetected neutral or ionized hydrogen are expected to be present around and between galaxies within a wall. This gas may be drawn from smaller galaxies or be the leftover from the two colliding shells that gave birth to the galaxies.

When velocities V_K and V_c converge, as observed at about ¼ and ½ of the curves, resulting forces bring masses closer to a circular pattern and may produce a density increment or the equivalent of a traffic jam, where new stars may be produced. But when velocities are equal, such as observed at the crossover in figure 6.23, masses are locked in and moving about a center of mass, which coincides with the center of attraction, so a dense circular or elliptical pattern may start to form. Finally, since ΔV is negative when V_c is smaller than V_K as observed near the disk's center, the mass associated with V_c may tend to diverge from the center and thus spiral out in the opposite direction to form an elongated bulge or the birth of a bar in NGC 3198.

Figure 6.24 Apparent mass, M_a, on the top and corrected mass, M, at the bottom for NGC 3198 using a 30-disk model.

The horizontal scale of figure 6.24 is the same as on the previous figure, and the vertical scale represents mass in solar masses (sun=1.99x10^{30} kg).

The top curve represents apparent mass, M_a, which is lower than the corrected mass for the first three values but increases generally faster than the corrected mass in this real example. It shows a sharp break to lower values at the edge of the disk but remains larger than corrected mass while converging. Apparent mass values reach a maximum of 153 billion solar masses and thus are 1.33 times larger than the corrected mass with a maximum of 115 billion solar masses at a radius of 30 kpc. Apparent mass is computed with the circular velocity equation 6.26, $Ma=V_c^2R/G$, where G is the gravitational constant, R is the radial distance, and V_c is the velocity measured from Doppler shifts or from matrix E. Apparent mass is not correct since it shows an impossible mass decrease away from the edge of the disk. As mentioned above, mass is positive and additive within a sphere or a disk and can only increase away from the center of a body or set of bodies. Note that in the big bang model, dark matter halos have been invented to maintain gravitational equilibrium in spiral galaxies because they use center of mass radiuses which are too high and uncorrected for gravitational disk effects and thus lead to excessive galactic mass evaluations and distributions.

The bottom curve increases from 0 to 30 kpc, and since no mass is added outside the visible disk, it is flat away from the disk over the next 10 kpc. It represents the cumulative calculated mass M from velocity curves V_c or V_K in figure 6.23 and R_c or R distance, respectively. This mass curve corresponds to corrected or real mass for galaxy NGC 3198. The circular velocity equation 6.24 or an equivalent may be used to compute the bottom curve M with $M=V_c^2R_c/G$, where G is the gravitational constant, M is the corrected mass, V_c is corrected velocity, and R_c is corrected distance.

Figure 6.25 Three curves from NGC 3198 corrected or real mass with central spheroid, no spheroid, and no spheroid with 2 kpc intervals.

Three different thin curves of the corrected or real mass are presented in figure 6.25 to test the sensitivity of the model to the presence or not of a central spheroid and to computation intervals of 2 kpc instead of 1 kpc with no spheroid. Central spheroid model is calculated every 1 kpc with a stack of 29 disks plus the first 1 kpc evaluated with classical Kepler-Newton's spherical model to account for the bulge at the center. The curve without a spheroid, almost indistinguishable from the first one, is computed every 1 kpc, exclusively with the complete stack of 30 disks. The third curve represents rougher values computed every 2 kpc with a stack of 15 disks and without a central spheroid. Total mass difference computed in these three ways amount to a maximum variation of about 3% for the 2 kpc intervals at the edge of the galactic disk. NGC 3198 corrected mass is thus evaluated at 115 billion solar masses. Thus, the method used in this book gives stable and reliable results.

Two papers published by Gallo and Feng [84] on galactic rotation with a thin-disk model similar to the above give a close value of 99 billion solar masses for NGC 3198 and 140 billion solar masses for the Milky Way, also close to our value of 160 billion solar masses. They used integrals to accomplish the summation in their model and found concordant results, including the conclusion that dark matter is repelled and not necessary to explain the galactic dynamics.

Figure 6.26 Mass added and density per ring for NGC 3198

The horizontal scale is restricted to the 30 kpc radius of the galactic disk, and the vertical scale represents solar masses on the left and density in M_p/m^3 on the right.

NGC 3198 is a large galaxy where the first five rings with 1-kpc-width contribute one and a half to over six billion solar masses. It is followed by a gradual decrease of added mass per ring since the average outer arms become more and more tenuous. The central ring is made up of a stack of the 30 central rings from the 30 disks and contains about one and a half billion solar masses, which include a most likely central black hole within the core.

Density is presented as M_p/m^3 per 1-kpc-width ring with a maximum average of 3 million M_p/m^3 in the central ring that contains a central black hole. A low density is reached near 500,000 M_p/m^3 at about ⅔ from the center. The density increase found toward the edge of NGC 3198 and many large galaxies indicates the merging of additional material from neighboring gas or galaxies. Now, any astrophysicist could combine this kind of results to multi-frequency survey to reconstitute a galactic distribution.

The same exercise will be applied to the small galaxy NGC 4062 to discover its force profile, velocity, and mass distribution.

6.2.2 Example from a Small Galaxy: NGC 4062

William Herschel discovered this small galaxy in 1787 in the Ursa Major constellation. It is classified as a small-nucleus Sc galaxy or as a small barred SBc galaxy with a radius of about 4 kpc and has a flocculent appearance.

Figure 6.27 Newton and corrected force for NGC 4062

No crossover is observed between Newton's force and corrected force values shown respectively as squares and diamonds in figure 6.27. The large hump associated with the corrected force is related to the presence of a large mass within the disk at a distance between about ½ to ¾ from the center and decreasing up to the edge. This may indicate the recent merging of some neighboring material that is in the process of integration to the galaxy.

Figure 6.28 Velocity distributions for NGC 4062

The velocity distributions of NGC 4062 in figure 6.28 do not show a crossover and are concordant with force distributions. Doppler or corrected velocities (square symbols) indicate a sharp increase to 166 km/s between ½ and ¾ of the radius, followed by a decrease to the edge of the disk at about 156 km/s. Finally, the Keplerian velocity (triangles) with a maximum of 139 km/s always shows the smoothest curve with the lowest values since it is based on the Newtonian concept that the whole mass is like a point at the geometric center. Since V_c is larger than V_K for the whole galaxy, then ΔV is positive; and since all orbiting material is spiraling in toward the center, no bar is expected.

Figure 6.29 Apparent and corrected mass for NGC 4062

The apparent and corrected mass distribution is plotted in figure 6.29. As expected from force and velocity variations, we find a sharp increase in the apparent mass at ½ to ¾ of the radius that is also present in the corrected mass. Apparent mass reaches a maximum of 22.7 billion solar masses, but the decrease away from the edge of the disk is impossible and indicates that apparent mass is incorrect. Corrected mass shows a constant mass value of 16.2 billion solar masses at the edge and away from the galaxy. Thus, apparent mass is an impressive 1.4 times more than corrected mass. This is a large error in mass evaluation typical of big bang model calculations based on center of mass radius. Dark matter is not justified, especially if we take into account the presence of poorly visible low surface brightness satellites and low-resolution neighboring clouds not accessible in the range of frequencies most commonly used in astronomy.

Figure 6.30 Mass added and density per ring for NGC 4062

The horizontal scale is restricted to the 4 kpc radius of the galactic disk in steps of 0.5 kpc for the rings, and the vertical scale represents solar masses on the left and density in M_p/m^3 on the right.

NGC 4062 is a small galaxy where the first five 0.5-kpc-width rings contribute from less than one to nearly three billion solar masses. It is followed by a rapid decrease of added mass per ring since the average outer arms become more and more tenuous. The central ring is made up of a stack of the eight central rings from the eight disks and contains about 900 million solar masses which likely involve a relatively small central black hole within the bulge. The almost-symmetric curve reversal is typical of smaller galaxies.

Density is presented as M_p/m^3 per 0.5-kpc-width ring with a maximum average of over 60 million M_p/m^3 in the central ring that may contain a central black hole. The lowest density around 30 million M_p/m^3 is reached in the last ring at the edge of the disk. The gradually decreasing density curve with flattening, and often curling back upward near the edge, is also typical of small galaxies.

6.2.3 Application to a Barred Galaxy: NGC 4389

Galaxy NGC 4389 is located in the constellation Canes Venatici and has an apparent magnitude between 11.7 and 13. It was discovered by William Herschel in 1788 and is classified as a small to intermediate barred spiral a SBbc galaxy. We include this galaxy as a model for a typical barred galaxy since our method involves the stacking of 12 disks for this galaxy, although a special treatment may be necessary due to the fact that a bar is not a spheroid or a disk but an elongated structure with some specific characteristics.

Figure 6.31 Newton and corrected force for NGC 4389

A strong crossover, located at a distance of 2.5 kpc (about 40%) from the center, is observed between Newton's force (squares) and corrected force (diamonds) in figure 6.31. The crossover is located near the edge of the bar. The slope of the corrected force is very steep between the center and the crossover indicating the presence of a bar. It rises with still a sharp slope to the edge of the galaxy, which points to the presence of an important mass spread the way to the edge at 6 kpc.

Figure 6.32 Velocity distributions for NGC 4389

The velocity distribution of NGC 4389 in figure 6.32 shows a crossover, which is concordant with force distributions. Doppler or corrected velocities (square symbols) indicate a two-step increase to the maximum at 115 km/s. Finally, the Keplerian velocity (triangles) with a maximum of 90 km/s shows a smooth curve since it is based on the Newtonian concept of the point mass at the center. The crossover represents a critical point where all the two velocity curves have the same value. Between the center and the crossover, we find that the Doppler velocity, when combined with center of mass radiuses, will undervalue masses since it is less than the Keplerian one, and from the crossover to the edge, Doppler velocity will overestimate masses. As expected, and to keep a constant angular velocity between the bulge and the crossover, both Doppler and Keplerian velocities increase approximately as the radius across the bar.

To the right of the crossover and up to the edge, ΔV is positive since V_c is larger than V_K, and consequently orbiting masses will spiral in at a velocity ΔV toward the center of the disk. When the vector ΔV is equal to zero at the crossover, then orbiting masses will form a circle or an ellipse around the center of the disk. To the left of the crossover, ΔV is negative since V_c is smaller than V_K, which means that orbiting masses will spiral out in the opposite direction at a velocity ΔV away from the center of the disk and build up an elongated bulge followed by a bar.

Figure 6.33 Apparent and corrected mass for NGC 4389

Figure 6.33 shows apparent and corrected mass distribution. As expected from force and Doppler velocity variations, we find a sharp increase in the apparent mass between the crossover and the galactic edge, while the corrected mass curve shows a more steady increase even before the crossover. Apparent mass reaches a maximum of 18.5 billion solar masses, but the decrease away from the edge of the disk is impossible and indicates that apparent mass is incorrect. Corrected mass shows a constant mass value of 11.1 billion solar masses at the edge and away from the galaxy. Thus, apparent mass is an impressive 1.67 times more than corrected mass. This is a major error in mass evaluation, especially when we consider the influence on the total mass of a galaxy from other factors such as a poorly visible galactic and gas material in the neighborhood that is not yet accessible with present-day technology.

Figure 6.34 Mass added and density per ring for NGC 4389

The horizontal scale is restricted to the 6 kpc radius of the galactic disk in steps of 0.5 kpc for the rings, and the vertical scale represents solar masses on the left and density in M_p/m^3 on the right.

NGC 4389 is a barred galaxy where the first nine 0.5-kpc-width rings contribute from 100 to 1350 million solar masses. It is followed by a decrease of added mass per ring since the average outer arms become more tenuous. The central ring is made up of a stack of the 12 central rings from the 12 disks and contains only 100 million solar masses which may still involve a small central black hole within the bulge. The asymmetric curve reversal is typical of mass increase toward the edge of the disk in barred galaxies.

Density is presented as M_p/m^3 per 0.5-kpc-width ring with a maximum near 19 million M_p/m^3 at the edge of the disk. The lowest density around 4.7 million M_p/m^3 is reached in the central ring that may contain a small central black hole. The gradually increasing density curve is typical of some barred galaxies where the average mass is shifted toward the edge.

6.2.4 Example from Elliptical Galaxy NGC 3379

Figure 6.35 Black and white version of elliptical galaxy 3379
(Credit: NASA/HST)

Elliptical galaxy NGC 3379 was selected to verify how much a disk model could apply to an apparent spheroidal-shape galaxy. Since our mass computation model involves the stacking of many disks of various sizes, it should yield an acceptable result for the mass of a spheroidal elliptical galaxy. NGC 3379 was discovered in the Leo constellation by Pierre Méchain in1781. It is classified as an elliptical galaxy E1, but since, a disk, or a circle of dusty-star-forming material is seen on some published pictures, as observed in figure 6.35, it could also be a S0 galaxy. In any case, the disk model applies since our method distributes the mass in a large number of disks of various sizes which may also reconstitute an elliptical galaxy.

Figure 6.36 Newton and corrected force for NGC 3379

Both curves from Newton's force (squares) and from the corrected force (diamonds) follow a similar Newtonian decrease away from the center in figure 6.36, betraying the presence of an extremely large mass near the center of NGC 3379. No crossover is occurring, but values from the corrected force clearly dominate the distribution and indicate the presence of some flattening likely intermediate between disk and spheroid models.

Figure 6.37 Velocity distributions for NGC 3379

Since no Doppler velocity flattening is present toward the edge, we find a more typical Keplerian velocity curve decreasing from center to edge. No argument may be invoked for dark matter with such a curve. Now, it is quite clear that going from a spheroidal to a disk model, the Doppler shift velocity curve will naturally go from a spheroidal decrease of velocities across the disk to a flattening of disk velocities toward the edge, or even increase in the extreme case of a homogeneous disk. In a spheroidal model, center of mass radius may be used as an approximation but, in a disk model, center of attraction must be used to find force distributions and masses.

No crossover is present in the velocity distribution of NGC 3379 in figure 6.37, which implies that ΔV is positive since V_c is larger than V_K across the distribution, and consequently orbiting masses will spiral in at a velocity ΔV toward the center of the galaxy. Contrary to the typical small galaxy NGC 4062 and to the barred galaxy NGC 4389 which reached their maximum velocities toward the edge of the disk, we find a maximum close to the center for NGC 3379, which makes it similar to the large galaxy NGC 3198. But NGC 3379 shows a large Doppler velocity excess near the center and tapering down toward the edge, while NGC 3198 has the large excess close to the edge. Thus, we expect, for NGC 3379, a smaller mass contrast between Doppler and corrected velocities, less than for the large galaxy NGC 3198.

Figure 6.38 Apparent and corrected mass for NGC 3379

Figure 6.38 shows apparent and corrected mass distribution for NGC 3379. As expected from force and Doppler velocity variations, we find a more rapid increase in the apparent mass near the center and much less added mass toward the edge, while the corrected mass curve shows a more steady increase. Apparent mass reaches a maximum of 101 billion solar masses at 12 kpc, but the decrease away from the edge of the disk is also impossible and indicates that apparent mass is incorrect. Corrected mass shows a constant mass value of 85 billion solar masses at the edge and away from the galaxy. Thus, apparent mass is 1.19 times more important than corrected mass. This is a remarkable error in mass evaluation, especially when we realize that we are dealing with an elliptical or possibly an S0 galaxy, where an intermediate model between disk and spheroid may apply.

Figure 6.39 Mass added and density per ring for NGC 3379

The horizontal scale is restricted to the 12 kpc radius of the galactic disk in steps of 1 kpc, and the vertical scale represents solar masses on the left and density in M_p/m^3 on the right.

NGC 3379 is an elliptical galaxy where the first two 1-kpc-width rings contribute 15 and 17 billion solar masses. It is followed by a gradual decrease of added mass per ring since the average outer arms or shells become more and more tenuous. The central ring is made up of a stack of the 12 central rings from the 12 disks and contains about 15 billion solar masses, which include a very likely massive central black hole within a near spherical bulge.

Density is presented as M_p/m^3 per 1-kpc-width ring with a maximum average of 82 million M_p/m^3 for the central ring that contains the central black hole. A low density is reached around 3.4 million M_p/m^3 near the edge. Both curves are similar to those of the large galaxy NGC 3198 where the central mass is predominant.

6.3 Statistical Results with 46 Galaxies

What would be the outcome if we apply our method and model on a larger number of galaxies with some statistical significance? To answer this question, 46 galaxies were selected from various published data, of which 16 are low-surface-brightness galaxies (LSB) [32] and 30 are high-surface-brightness galaxies (HSB) [29] [38] [47].

Column 1	Column 2	Column 3	Column 4	Column 5
HSB galaxies:			LSB galaxies:	
Name of Galaxy	Galactic radius in kpc	Comments	Name of galaxy	Galactic radius in kpc
NGC 1003	1.0		DDO 64	2.5
NGC 4096	1.5		DDO 168	4.0
NGC 4448	2.0		UGC 1281	5.0
NGC 4062	4.0	Small selected	NGC 2366	5.5
NGC 2708	4.5		NGC 4455	6.0
NGC 3495	5.0		F 583-4	6.5
NGC 4389	6.0	Barred selected	UGC 731	7.0
NGC 3691	7.5		DDO 170	7.5
NGC 2259	8.0		NGC 100	8.0
NGC 3949	9.0		DDO 154	8.5
NGC 1417	10.0		F 568-3	11.0
NGC 3379	12.0	Elliptical selected	UGC 4173	12.0
NGC 4527	13.0		F 571-8	14.0
NGC 3917	15.0		UGC 711	16.0
NGC 2590	16.0		UGC 3137	26.0
NGC 4217	17.0		UGC 1230	36.0
NGC 2403	19.0			
NGC 3079	20.0			
NGC 3031	21.0			
NGC 4088	22.0			
NGC 3521	23.0			
NGC 2903	25.0			
NGC 6946	29.5			
NGC 3198	30.0	Large selected		
NGC 3992	32.0			
NGC 3726	34.0			
NGC 7331	36.0			
NGC 224	38.0			
NGC 5055	40.0			
NGC 2841	44.0			

Table 6.1 List of 30 HSB and 16 LSB galaxies with radius in kpc

In column 1 of table 6.1, we find the list of 30 HSB galaxies with their respective NGC denomination, followed in column 2 by their maximum radius in kiloparsec (kpc). As found in literature, both HSB and LSB galaxies are related to Doppler redshift signatures that have been corrected and converted to rotation velocities. Average Doppler velocities are sampled and extracted from these rotation velocities in order to compute corrected and Keplerian velocities, masses, densities and periods. Column 3 indicates the four selected galaxies that were previously presented as typical galaxies. The list of LSB galaxies is found in column 4 with their respective maximum radius listed in column 5.

Figure 6.40 Corrected galactic mass at the edge of selected galaxies

The corrected galactic mass at the edge of 46 selected galaxies is presented in figure 6.40 with one common horizontal scale and two different vertical scales. The horizontal scale is from 0 to 50 kpc, and the two vertical scales are in solar masses, but the top one is a logarithmic scale to show the better resolution for galaxies with a smaller radius. The black diamond symbols represent HSB galaxies, while the gray squares represent LSB galaxies. We observe that corrected galactic mass increases with the radius of galaxies, and we expect this increase to be in proportion to the square of the radius for a uniform disk. But the mass is increasing with less than the square of the radius since mass is not uniform in a galaxy and decreases generally and gradually from the massive bulge at the center.

The most important result is that the mass of LSB galaxies are lower than the mass of HSB galaxies all the way across the distribution. Having less mass and thus less star-forming clouds and regions seems a perfect reason for a galaxy to exhibit low surface brightness.

Figure 6.41 Ratio between apparent and corrected mass for 46 galaxies

Ratios between apparent and corrected masses are presented in figure 6.41 for the 46 selected galaxies. The vertical scale shows that all ratios are found between 1.1 and 1.7. Both HSB and LSB galaxies follow the same trend, with decreasing apparent-to-corrected-mass ratios across the distribution from about 1.5 to 1.3. The curve indicates larger ratios for small galaxies and seems to be flattening at larger radiuses. The position of four galaxies discussed above, as well as Andromeda, is indicated as a reference. We note that NGC 4389, the barred galaxy, has the highest apparent-to-corrected-mass ratio at 1.67, perhaps indicating that large barred galaxies may contain the largest errors in mass estimates or the need to improve the model when a bar is present. On the other hand, the elliptic galaxy NGC 3379 shows a low apparent-to-corrected-mass ratio, perhaps indicating a drift from the disk model to a more spheroidal model.

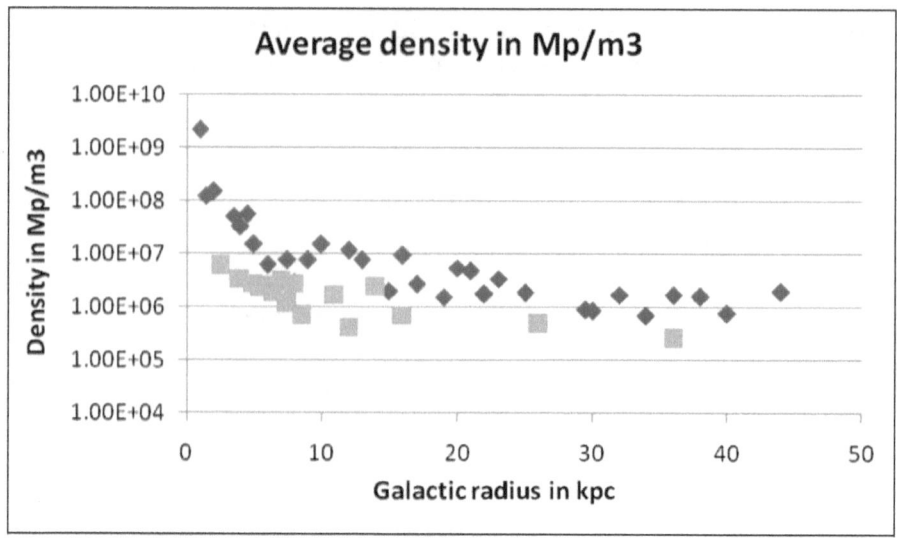

Figure 6.42 Average density in M_p/m^3 for the 46 selected galaxies

The average density across the disk of 46 selected galaxies is presented in figure 6.42 with a horizontal scale from 0 to 50 kpc and a vertical scale in M_p/m^3, with logarithmic divisions to show a better resolution for galaxies with a smaller radius. The black diamond symbols represent HSB galaxies, while the gray squares represent LSB galaxies.

We observe the surprising and unexpected result that small radius galaxies have a higher density and that density generally decreases with increasing galactic radius for both HSB and LSB galaxies. The decrease is faster in the first 6 kpc from the center and then slower to 25 kpc from where it flattens out to the edge. Density varies from a maximum 1.3 billion M_p/m^3 for the smallest HSB galaxy to a minimum near 360,000 M_p/m^3 for the largest LSB galaxy. These are averages across the disks since density is not uniform within a galactic disk and decreases generally and gradually from the massive bulge at the center, except for some barred galaxies in which density increases between the center and the edge.

But how can we explain that the smallest galaxies have the largest densities? Two hypotheses may be presented: the first model would imply that small galaxies were originally large galaxies that have been stripped of their gaseous arms, leaving only the denser portion of the inner disk; the second model implies that the material from the original arms of the smaller galaxies has been concentrated toward the center by spiraling in, thus increasing the density. The first model may predominate in areas where galaxies are in the process of merging, and the second model may apply to isolated and starved galaxies.

Another important result is that the density of LSB galaxies is lower than the density of HSB galaxies all the way across the distribution. Having less density implies less mass and, as above, less star-forming clouds and regions, which correlate with low surface brightness for a galaxy.

Figure 6.43 Average spiraling in velocity for low- and high-surface-brightness galaxies

The vertical scale of velocity varies from 2 to about 28 km/s in figure 6.43. It represents the average spiraling in velocities for our 46 selected galaxies, where $\Delta V = V_c - V_K$ is the average of all ΔV values for each galaxy. Since the angle between vectors V_c and V_K is zero and since we are using averages that apply about halfway across galactic disk radius, we can use vector ΔV as a good approximation for the spiraling in velocities. Again, diamond symbols represent HSB galaxies and squares correspond to LSB galaxies.

We observe a remarkable difference between HSB and LSB galaxies, where LSB values are generally much lower than HSB delta velocities and represent a much slower spiraling in movement of orbiting bodies around LSB galaxies. HSB galaxy NGC 4389, our special barred example, is oddly right in the middle of LSB galaxies due to a strong crossover and negative values that will affect the ΔV average. Elliptical galaxy NGC 3379 shows a large average ΔV of about 20 km/s, along with a few other galaxies where a relatively large nucleus could also be present.

Figure 6.44 Average time to spiral in a distance equal to ½ the circumference at the galactic edge

In figure 6.44, we find the average time for a body to spiral in a distance equal to ½ the circumference at the edge of the galaxy. The vertical scale is time in years; diamond symbols are still related to HSB, and squares to LSB galaxies. Both LSB and HSB time values increase with the distance since time-distance is directly related to the radius, but there is a clear cut in the distribution of LSB and HSB galaxies whereby the spiraling in time taken by LSB galaxies is higher than for HSB galaxies. from 0.11 to 17.5 billion years for LSB galaxies and implies slower orbiting down the spiral arms, lower masses, and less activity as observed. For the HSB galaxies, time varies from 120 million to a maximum around 18 billion years, thus suggesting a faster mutation to ellipticals for these higher-energy galaxies.

6.4 Orbiting Mass and Spiral Galaxies

Spiral and disklike galaxies are complex gravitational systems with massive cores surrounded by a bulge that may extend to a bar before the spreading of the arms, followed by ill-defined nebulous material related to satellites and to the surrounding wall. A sketch is drawn in figure 6.45 to help in understanding how a material may be spiraling in from the outside to the inside of a galaxy and to show how, from mass-to-mass interconnections, centers of attraction may migrate to the center of mass. The near center spiraling out is neglected here.

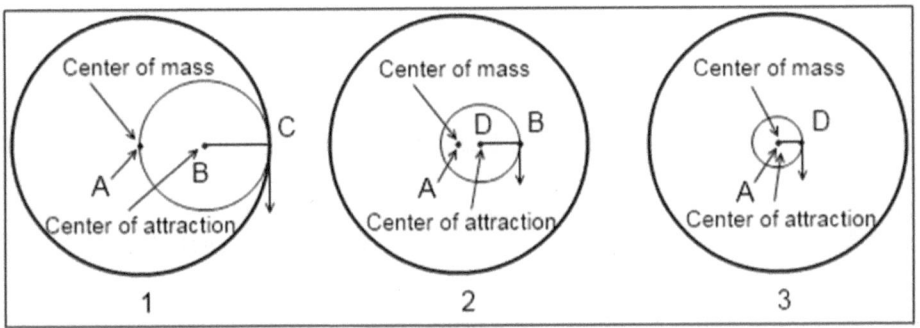

Figure 6.45 Migration of the center of attraction to the center of mass

Three circles representing the disk of a galaxy are drawn to shown how the center of attraction of different masses is gravitationally interconnected to the center of mass of the whole disk located at A, which is also the geometric center.

On disk number 1 at the left, we have an orbiting mass at C, whose center of attraction is located at B, halfway between A and C, so that the mass at C is moving with the Doppler or corrected velocity, V_c, in the direction of the arrow and around the small circle because of the centripetal acceleration. Note that the center of attraction B is like a barycenter since for the mass at C, it is gravitationally acting as if the whole mass of the disk were concentrated at that point.

But on disk number 2, at the center of the figure, the mass that was located at the center of attraction at B in disk number 1 is now the orbiting mass around the center of attraction located at D so that mass B is orbiting with the corrected velocity, V_c, around the small circle. Note that as we go from orbiting masses at the edge to orbiting masses toward the center, the small circles decrease gradually in radius with the consequence that tangential movements draw a stronger and stronger curvature corresponding to a spiral.

Finally, on disk number 3 at the right, the mass that was located at the center of attraction at D in disk number 2 is now also the orbiting mass around the center of attraction located at A so that mass D is orbiting the Doppler or corrected velocity, V_c, around the small circle, and both the center of attraction and the center of mass are located at the same point A, which is also the geometric center. Since the displacements at C, B, and D are gravitationally connected and occur simultaneously, they participate in the overall rotation of the galactic disk around the center of mass and correspond to the Keplerian velocity, V_K.

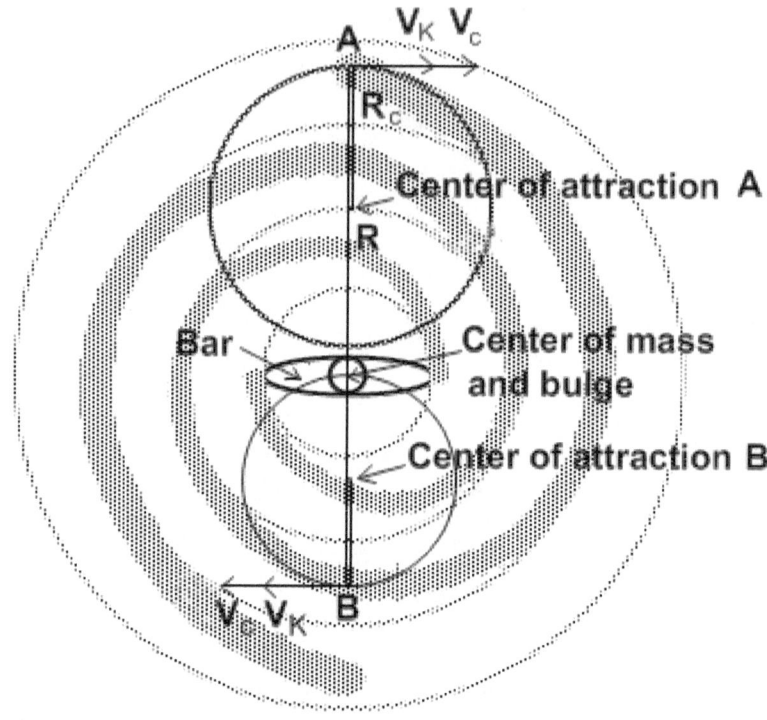

Figure 6.46 Sketch showing the relationship between orbiting mass and a galaxy

A galaxy is drawn with central core, spheroid bulge, bar, and two gray-shaded spiral arms in figure 6.46. Four concentric circles are shown at equal intervals around the center of mass as a background reference. The bulge is the spheroidal shape near the center of the galaxy, where Kepler-Newton laws may approximately be applied. The bar is that part of a galaxy where rotation velocity between the bulge and arms is equal to angular galactic velocity. Tangential velocity increases in the bar as the radius R. The bar is fed in fresh mass by the arms of the galaxy, and the oblate spheroid core, in turn, may be fed by the bar that should eventually spiral down to the bulge and core in a starved situation.

The mass present at A is associated with two velocity vectors whose lengths are proportional to velocity values. Keplerian velocity, V_K, is a spherical equivalent associated with center of mass radius $R_K = R$. Corrected or Doppler velocity V_c, is associated with radius R_c running from A to center of attraction A.

A circle is shown around center of attraction A with radius R_c. This circle is much smaller than the center of mass circle (not drawn) passing through orbiting mass A. But the orbit of mass A is deflected by the stronger curvature associated with velocity V_c while rotating around the larger center of mass circle with velocity V_K.

The same observations apply to mass B, where a still-smaller circle is shown about 180 degrees down the spiral. The latter smaller circle implies a yet-stronger spiral curvature. Therefore, gravitating masses around gradually smaller circles with smaller corrected radius R_c will form a progressively stronger spiral curvature toward the galactic bar or central spheroid. But other significant compensating forces may be involved, like tidal effects creating a drag that could disturb the orbiting path and that may unspool the galactic arms of smaller galaxies during the merger with a larger companion. These effects will produce spiral arms curving gradually more and more and funneling masses toward the center of the galaxy. So the spirals, in more stable galaxies, run from outside to inside and may be expressed as a decreasing function of radius R, some of them looking like a double Fibonacci spiral.

Figure 6.47 Distribution of orbiting time and spiraling in time for NGC 3198

The distribution of orbiting time for one turn (diamonds) and spiraling in time (squares) for NGC 3198 is shown in figure 6.47. The horizontal scale is the radius in kpc, and the vertical scale is time in years. Orbiting time of 300 million years for one turn at 8 kpc is indicated as a comparison to our sun orbiting the Milky Way galaxy at about the same distance in a shorter time of 250 million years, so NGC 3198 is a less massive galaxy with a larger radius (30 kpc versus 20 kpc) and with a small bar. The crossover corresponds to a circular low delta V velocity between the bulge-bar and the arms when center of attraction and center of mass coincide, and the two negative time values on the left of the crossover represent the absolute time value for a spiraling out in the opposite direction away from the center which corresponds to an elongated bulge or the beginning of a bar. Orbiting time goes from 0 to 1.5 billion years, and, except at the crossover, spiraling in time for a distance equal to ½ the circumference of the ring varies mainly between 2.3 and 9.5 billion years. For example, in NGC 3198, at a radius of 24 kpc, a star takes one billion years to do a complete orbit around the galaxy and 6.2 billion years to go down the spiral toward the center over a distance of about 65 kpc, which is equal to ½ the circumference at 21 kpc.

Except near the edge, the most outstanding relationship is that it takes the same number of turns (about eight) to spiral in the same angle of 180° across most of the galaxy so that the same spiral shape is preserved for many billion years, while new material is fed from the outside, and the gravitating material of the arms is incorporated to the galactic bulge. But in the last third of NGC 3198 near the edge of the disk, where spiraling in takes less than eight galactic turns for the proportional half-circumference distance, material should be falling in faster and stimulating star formation.

Orbiting masses with different radial values are not always moving at the same angular velocity in galactic arms since a galaxy is not a rigid disk. Angular velocity decreases with increasing R distance when a flat tangential apparent velocity curve is present. Therefore, an orbiting mass near the edge of a galaxy will take a longer time to go down its spiral path (billions of years); and conversely, a mass near the center will spiral in faster.

Most masses are rotating with a small differential velocity proportional to $\Delta V = V_c - V_K$, which represents, for NGC 3198, an average across the galactic radius of 5.4 billion years to go down a spiral equal to ½ the length of a circumference with a 15 kpc radius and for an average $\Delta V = 8.6$ km/s. Our sun, for example, is spiraling in toward the center of the Milky Way galaxy and should take more than two billion years to reach the bar down the spiral at a distance of about 25 kpc with a $\Delta V = 15.5$ km/s. By that time, the sun should be emitting longer wavelengths in the red and swell to become a red giant star, and the Milky Way would be half way closer to merging with the larger Andromeda galaxy in our local group.

This indicates that most orbiting masses in every spiral galaxy are gradually and irremediably spiraling down toward the center. Thus, older stars must be found near the galactic center. So spiral galaxies seem to sweep and gobble down all material from their surroundings over billions of years until exhaustion or merger. Moreover, since the outer arms of spiral galaxies have a larger cross-section than the inner portion of the spiral, the material is not only spiraling in but also funneling in and concentrated in denser clouds where molecular hydrogen forms and dynamically generates stars.

A generalization of the basic model applied to nonsymmetrical fields of other celestial bodies should give a simple and more rational explanation of the advance of Mercury's perihelion since our sun is an oblate spheroid and has a measured flattening of about 6 km at each pole, which seems the right order of magnitude to produce the small advance in the planet's orbit. In other words, the perihelion advance should correspond to a spiraling in of Mercury toward the center of the sun in the same way that stars spiral in toward the center of galaxies. The same difference between Newtonian and corrected force should apply to calculate proper center of mass and center of attraction of any non-perfectly spherical system; and although to a much lesser extent, to any planet-moon system or any multiple-star system since some flattening is generally present.

6.5 Linear Model for Filamentary Structures in Galaxies, Groups, and Clusters

Interacting and merging galaxies exhibit long filamentary gas and star trails captured by their tidal influence on smaller galaxies and clouds. Galactic groups, clusters, and superclusters apparently exhibit similar mass-to-light (M/L) ratio and velocity discrepancies as discussed for single spiral galaxies but with a much larger factor. Note that a supercluster is generally the apparent assemblage of two or more clusters near the same line of sight before any low redshift, neutral shift, or blueshift corrections have been applied in 3-D.

Figure 6.48 Linear model used for computations

The linear 1-D model of figure 6.48 shows a test mass of one unit located at the center of a 0 to 300 distance units line, which means at 150 units from both ends. So these are the same distances as the center and diameter of the 2-D disk model presented above. A sequence of 150 masses of one unit, all separated from the next one by a distance of one unit, is present on each side of the test mass. The test mass is moved in steps of five units to one end of the line, up to 150 units, for a total of 31 steps that are analyzed in a way similar to the larger disk of 150 units in the previous disk model but, in this case, without an extension of 50 units away from the end of the line.

Figure 6.49 Corrected and Newtonian forces for a linear mass distribution

The horizontal scale runs from five to 150 units in steps of five units in figure 6.49, and no values are plotted for distance 0 since the vertical scale is logarithmic and cannot display zero. Gravitational force computations F_c are done for each separate mass and distance unit to obtain the progressively cumulative ΣF_c. Force F_n is computed from the center of mass as the linear mass and distance increase on one side and decrease on the other. From these, as for the previous disk model and with the same equations, masses and velocities will be calculated for an orbiting mass close to the line.

Figure 6.50 Corrected velocity, V_c (squares); and Keplerian velocity, V_K (triangles).

In figure 6.50, the horizontal scale also runs from five to 150 distance units in steps of five units. The test mass is moved from the center at the left and across the linear mass distribution toward the right. Vertical scale shows velocity in nonspecific units, with zero absent on the logarithmic scale. We recognize a similar pattern of velocity increase with crossovers as with the previous disk model in figure 6.13.

The curve with squares, ranging within two orders of magnitude and crossing the other curve, corresponds to the equivalent of a Doppler corrected velocity, V_c, which, when combined with center of mass radiuses, will give undervalued masses on the left toward the center and overvalued masses toward the right end of the line. This is the kind of large velocity variation that could be measured from Doppler shifts in, for example, bars, linear galactic arms, linear strings of galaxies, or elongated groups and clusters of galaxies.

Squares represent corrected velocity V_c that will yield corrected mass when combined with corrected distance R_c in the circular velocity equation. The flat curve with triangle symbols corresponds to Keplerian velocity V_K that will also yield corrected mass when combined with center-of-mass distance R in the circular velocity equation.

Figure 6.51 Distribution of apparent and corrected mass

The horizontal scale is the same as on the two previous figures, and the vertical scale with the 0 excluded is logarithmic and represents mass in nonspecific units in figure 6.51. The crossover at about 110 units separates the most significant and overvalued apparent masses to the right from the least significant and undervalued apparent masses to the left and this all the way to zero. Apparent mass is 11 times larger than corrected mass at the end of the line. Corrected or real mass M (squares) varies from zero to 300 mass units as selected in the original model in figure 6.48, but the first value shown on the log scale is 10 (2X5) since both sides of the line are added.

Figure 6.52 Ratio apparent to corrected or Keplerian mass M_a/M.

The horizontal scale in figure 6.52 is the same as the previous figure and starts at a distance of five units. Vertical logarithmic scale indicates ratios between apparent and corrected mass M_a/M. The force and mass crossover is indicated where Ma=M at 110 distance units from the center of the line.

Thus, the curve represents ratio increases from the center to the end of the line between apparent mass M_a and corrected or Keplerian mass M, where the latter two correspond to real mass M. The increase is impressive since apparent mass varies about four orders of magnitude across a linear distance of 145 units from five units to the end. It means that mass could be underestimated across the least important orders of magnitude near the center of such a linear pattern and overestimated across the most important orders of magnitude toward the end of such a linear pattern.

For this linear model, we have an apparent mass 11 times larger than the corrected mass. This ratio is much more important than that in the previous disk model, where the apparent mass was 2.16 times larger than the corrected mass for the theoretical model, and on average, Ma/M was 1.38 times larger for the 46 selected galaxies. If the same difference is present for the linear model between theoretical and applied mass results, then we should find that apparent mass is 8 times larger than corrected mass on average.

Since bars, filamentary structures, and some groups and clusters are generally described as more or less elongated features, it should be no surprise to find a large range of values for apparent-to-corrected-mass (M_a/M) ratios in field measurements, which would translate into a large range of mass-to-luminosity M/L ratios, much more than observed for disk-shaped individual spiral galaxies. This linear model may also be applied to some galaxies with elongated arms, such as NGC 6872 or UGC 10214, and in merging galaxies with trails of material. Finally, the linear model may be used to explain part of the excess velocities found in galaxies feeding clusters via the concave triangular ducts located between the walls and the clusters.

7. FOUR FORCES OR FACETS OF ENERGY IN THE COSMOS

Most books of physics and astrophysics state that four basic forces have been identified in the cosmos: gravitation, electromagnetism, the weak force, and the strong force. But the last three forces occur and are intertwined at the atomic level and may be grouped into a unique force—the electronuclear, which is described by the theoretical grand unification theory. So this would leave the two known basic forces, gravitation and the electronuclear, to which we may add a third one: the hypothetical black force that stops the collapse of massive black holes when all electronuclear forces have been exhausted. A fourth one may be suggested for the complexity bonds from subchemical to biochemical reactions. The complexity bond is this diffuse force driving reactions in the cosmos from the most initial and farthest electromagnetic ones to the most complex biochemical and from which stems survival and these associated cosmic qualities that we call cooperation and intelligence.

7.1 Electronuclear Force and Energy Conservation

The electronuclear force results from the interaction of all the fermions and bosons, as described in the table of fundamental particles in the Standard Model of physics. Since there are energy and particle exchanges at the quantum level and friction everywhere at the macroscale, thermodynamics dictates energy conservation and entropy in all processes.

7.1.1 Standard Model of Particles

The Standard Model of particles counts 12 fundamental elementary particles called fermions and five fundamental bosons, including the Higgs, as shown in figure 7.1. The figure represents an instant flash picture of the present-day knowledge of physicists that is continually moving forward to new confirmations and discoveries.

Fermions				Bosons			
1	2	3			spin 1		
6 quarks: spin 1/2							
2.4 MeV 2/3 up	1.27 GeV 2/3 charm	171.2 GeV 2/3 top	$m=E/c^2$ charge name				
4.8 MeV -1/3 down	104 MeV -1/3 strange	4.2 GeV -1/3 bottom			0 0 gluon		
6 leptons: spin 1/2							
0.511 MeV -1 electron	105.7 MeV -1 muon	1.777 GeV -1 tau		spin 0	0 0 photon		
< 2.2 eV 0 electron neutrino	< 0.17 MeV 0 muon neutrino	< 15.5 MeV 0 tau neutrino	125 GeV 0 Higgs boson	91.2 GeV 0 Z^0 boson	80.4 GeV +1 W^+ boson		80.4 GeV -1 W^- boson

Figure 7.1 Fermions and bosons of the Standard Model

The 12 fermions are subdivided in six quarks and six leptons with spin ½. Each of them is represented by a little square, where the first value on the top is the energy of the particle, which may be divided by c^2 to obtain the mass. It is followed below by the charge of the particle, which corresponds to uneven values of ⅔ and -⅓ for the quarks and to even values of -1 or 0 for the leptons. The name of the particle is given at the bottom of each square. There is a crescendo in the mass of the particles from the first generation in the first column on the left to the third column on the right. Most of the known cosmos is composed of the up-and-down quarks forming nuclei, which are combined with electrons to form atoms. The photon is the exchange boson that we are most familiar with, but the unfelt bosons are just as important in the nuclear reactions, together with the flow of neutrinos.

The five bosons are stranger since two of them, the photon and gluon, are considered massless and without a charge, and the other three, Z^0, W^+, and W^-, have masses of 91.2, 80.4, and 80.4 GeV and charges of 0, +1, and -1, respectively. The latter three are the exchange bosons for the weak force or interaction in beta and other nuclear decays and reactions.

Photons represent transient energy absorbed or emitted by particles both at the electronic and nuclear levels. Their energy and momentum depend only on frequency, and they are related by the equations E=hf =pc, where h is Planck's constant, f is frequency, p is momentum, and c is their velocity in a vacuum. Photons have no rest mass in the Standard Model; they carry the electromagnetic force and behave both like waves and elementary particles described as discrete quanta of energy.

Like photons, gluons are exchange particles carrying energy between different types of fermions at the nuclear level. They carry the strong nuclear force between quarks and have no rest mass in the Standard Model. No isolated gluon has ever been studied, and no quark either, since both are glued so strongly together with a force that increases with distances in the nucleus; no particle accelerator has succeeded yet to separate a single gluon by breaking this link.

The Higgs boson is an elementary particle under study by some of the teams at CERN near Geneva. It is an exchange particle that is proposed to explain the mass of most particles and the absence of mass for two of the bosons. In 2012, CERN scientists announced the discovery of a particle compatible with the Higgs boson, but much work is still needed to confirm the attributes of the particle.

Except for the photon considered as its own antiparticle, for every particle, there should be an antiparticle born in high-energy pair production, but that kind of symmetry obviously does not exist since the cosmos, as we experience it, is almost made of particles only. So somewhere in the process of particle production within the maxibang events, the symmetry is broken, and particles emerge as the dominant half of the pair. Matter has survived the antimatter annihilation because some unknown subparticle or subboson has intervened and broken the symmetry early in the process of bubble expansion. The experimental proof has not yet been demonstrated.

Figure 7.2 UCT quarks and projection to Planck's energy

Up, charm, and top (UCT) quarks and projection up to Planck's energy are shown in figure 7.2. The horizontal scale represents generation or family number, and the vertical scale is energy in eV with logarithmic divisions in multiples of three orders of magnitude.

The first three values of 0.0024, 1.27, and 171.2 GeV correspond to the three quarks with charge ⅔ in figure 7.1. The last value of 1.22×10^{19} GeV represents Planck's energy from the big bang model. It would be located at the gravity wall, where the electronuclear energy would not cross but would rebound if transposed to a maxibang model. All the squares in between are approximately distributed with a constant interval, and they represent the *mare incognitum* of particle physics. The first square corresponds to a level of energy of 45 TeV, while the Large Hadron Collider at CERN is expected to reach only 14 TeV in the year 2014.

Now imagine all the new particles and families of particles that could lurk in *mare incognitum*. Thus, it seems research and new discoveries await the future generations of humanoids. The task of scientific discoveries is far from being completed since we have no information covering most of the expected energy range in the cosmos. Beyond the known electronuclear, we should find a high-energy realm that will explain the early maxibang expansion at superluminal velocities and also a most likely new subquark force, the black force, that halts the collapse of black holes to within finite values, and this all the way to the instant before the maxibang implodes and explodes from the most massive black hole containing the equivalent mass of about seven thousand galaxies at the critical point or the equivalent of 171 trillion sun masses.

7.1.2 Friction, Entropy, and Energy Conservation

Tectonic plates creep and shake, wheels squeak, and atoms heat up when squeezed, so friction is present everywhere, and the pure mathematical laws of physics are plagued with imperfections because we live in a dynamic electronuclear domain of the cosmos where movement is sparked by the exchange of energy. So the four laws of thermodynamics apply equally to all electronuclear processes in expanding bubbles, in galaxies, in clusters, and in quasars. Of course, there is one exception with the gravitational implosion/explosion in maxibang events where a tiny portion of the cosmos is rewound by the gravitational energy, and entropy is set back to zero. But the whole cosmos is a closed system where the first law strictly applies: nothing is lost, nothing is created, and every bit of mass-energy is recycled.

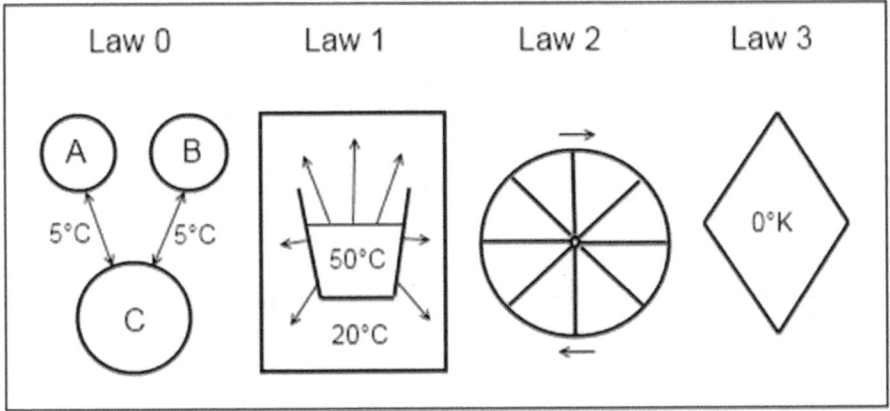

Figure 7.3 Sketch of the four laws of thermodynamics

Law number 0 of thermodynamics simply states that when two systems, A and B, are separately in thermal equilibrium with a third one, C, then they both are in thermal equilibrium with each other. Translated in terms of temperature, this means that we can measure and compare temperatures from different systems. This is currently used in astrophysics when radiations from stars or galaxies are compared to establish their level of activity.

The first law is the best known and easy to understand. It states that energy is conserved in all processes since nothing is lost, nothing is created, but everything is transformed. Thus, heat and work only have the effect of adding or removing energy from a closed system without affecting the total energy balance sheet. This law implies that a perpetual-motion machine is impossible since an input of energy is necessary to do some work in the electronuclear realm. In figure 7.3, a vessel containing a liquid at 50°C radiates its energy to the surrounding room at 20°C so that heat is only transferred to the room until thermal equilibrium is reached.

The second law is well-known and often cited but more difficult to grasp since it nearly always represents the increase of a negative factor in energy exchanges. The law states that the entropy of an isolated system never decreases but that it almost always increases so that such a system evolves toward a state of maximum entropy while reaching thermal equilibrium. Since converting energy to work without frictions or any other side effects is impossible, then building a perpetual-motion machine that would convert endlessly one source of energy to work is also impossible. The wheel on the figure is flawed with friction, so its kinetic energy is bound to decrease gradually while increasing entropy.

The third law deals with the entropy of a system at low temperature. It states that the residual entropy tends to a constant value as the temperature of the system tends toward zero, and thus, at the limit in the classical interpretation, the entropy of a pure crystal is zero at the absolute zero and represented as 0 on the Kelvin scale. The law is symbolized with a diamond at 0 K in figure 7.3.

7.2 Gravitation and Electronuclear Rewinding

As already mentioned, the force of gravity plays a unique role in the cosmic rewinding of all the electronuclear energy. Since gravity is a conservative force, it has special attributes not found in the electronic or nuclear forces. The speed of the gravitational interaction has never been measured, but it must be more than constant c since gravity is not confined within the black holes.

7.2.1 The Conservative Force

In classical physics, when the work done to move a particle from point A to point B is independent of the path followed, then the force is conservative. If the particle moves in a loop from A to B and back to the same kinetic energy in A, then the total work is zero for a conservative force. Gravity is a conservative force that depends only on the position of the object to which a potential value may be assigned. On the other hand, friction represents energy losses in the form of heat and is an example of a nonconservative force.

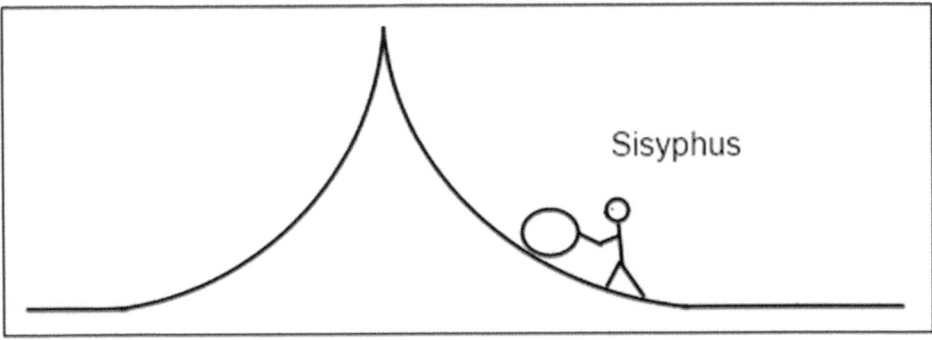

Figure 7.4 The fate of Sisyphus rolling forever a boulder up the hill

In Greek mythology, Hades was lonely as a god-king in the underworld across the river Styx. He abducted Persephone, the daughter of Zeus, and made her his wife, so they jointly welcomed their earthly human clients. Sisyphus, who had reigned as king of Corinth, was a new client for Hades but also a cunning artist who handcuffed the latter so that he could keep on partying on earth. This action created havoc in the earthly entropy since nobody could cross the Styx anymore. Hades was finally liberated, but Sisyphus tricked Persephone and borrowed an extra time of pleasure on earth. Such a crime against a god and a goddess was severely punished; he was condemned for eternity to roll a huge stone up to the top of a hill with a tremendous effort just to see the stone roll back to the bottom and then start all over again.

This story illustrates how the force of gravity does not rub off or wear out no matter how often Sisyphus repeats his absurd task. The potential gravitational energy is always present with the same intensity at the top of the hill, and the stone acquires kinetic energy as it rolls downhill while the potential energy decreases.

7.2.2 Gravity and Velocity

What is the velocity of gravitational interconnections between two units of energy in the cosmos? And what is the velocity of the effects of the gravitational force on the electronuclear particles? These are two of the most fundamental questions in our quest to understand the cosmos [50].

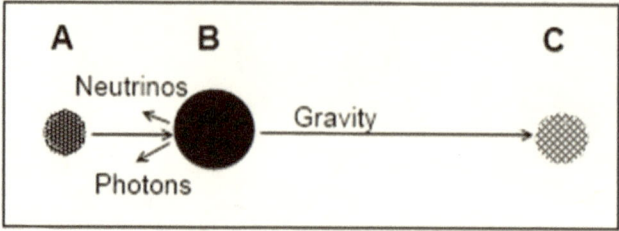

Figure 7.5 Gravitation is the deepest and most overpowering of all forces

Although gravitation is the most familiar and inescapable force of nature, it is also the most remote and out of reach with present-day technology. Figure 7.5 shows three bodies, A, B, and C, lined up and interacting. Electronuclear energy emerging from A will be absorbed or reflected by the massive neutron star located at B and so will also be neutrons and neutrinos. But the gravitational field influence of A is fully felt by every neutron in B and also by every particle in C. The presence of B does not alter or deviate in any way the gravitational influence of A on C. In other words, we could remove B, and the gravitational force between A and C would be exactly the same. No experiment has ever proven the contrary.

Moreover, there is a physically measurable delay for photons and neutrinos reaching B. The delay is proportional to the speed of light, c, for photons and slightly less for neutrinos, but no serious experiment supporting the propagation of gravity at light's speed has ever been demonstrated in the past, and no measurable aberration related to propagation delay may be observed for gravity between A and B or C. Gravity is the permanent physical contact between every two units of particle energy in the cosmos that obeys to the inverse square law in one dimension. But the effects of gravity on the electronuclear are measured and perceived at a maximum speed of c since photons emitted in such an interaction cannot travel faster than light.

The speed of gravity is not the speed of light or an infinite speed, but it is simply a permanent instantaneous-like contact in a cosmos with finite dimensions. In other words, two units of energy are always within finite reach of each other. Otherwise, it would be utter chaos in the cosmos without the cohesive force of gravity that glues together all particles within the 286 billion l-y diameter of the spheroid. For example, two electrons with a mass of 9.11×10^{-31} kg each and separated by the cosmic diametrical distance of 2.71×10^{24} m are gravitationally attracted to one another by an extremely weak force of 7.55×10^{-120} Newton. Over cosmic distances, forces and densities are so low that there is no risk of collapse, and bubble expansion is in equilibrium with gravity contraction.

The gravitational field notion covers the effects of a large number of energy units, like within a proton or a star. The theory of general relativity from Albert Einstein is a special description of the time-space behavior of electronuclear particles in a cosmos under the control of gravity. It is special relativity that brought the confusion about the speed of gravity by stating that nothing can exceed the speed of light. But general relativity uses instantaneous coordinates and momenta in its equations of motion, thus implicitly recognizing faster-than-light gravity. Theoretical physics has continued to strive toward a theory of everything, which would unify the Standard Model with quantum gravity and large-scale relativity [71] [79]. The most promising attempt seems to be loop quantum gravity, but as we have seen, the scale of energies is so tremendous that no machine will be able to probe to the core of gravity in the foreseeable future unless we find a way to extract useful information from the black holes.

7.2.3 Cosmic Cyclic Rewinding

We will review the time sequence computed and projected by big bang proponents and interpret this sequence as a maxibang event with its peculiarities. Then we will use Planck's length as the radius of a sphere, compute the volume, and find the density and other parameters for a compacted or expanding mass-energy equivalent of about seven thousand galaxies. Finally, we will briefly describe a heat pump and apply its physical characteristics to the cosmic self-rewinding heat pump.

Bang time sequence for our bubble				
Time	Time	Energy	Temperature	Event
	in seconds	in eV	in K	
13.8 billion years	4.35E+17	2.59E-04	2.73E+00	Present age of our bubble
1.0 billion years	3.16E+16	6.29E-03	1.00E+02	Galaxies form around our bubble
380 000 years	1.20E+13	2.59E-01	3.00E+03	Radiation starts to escape
70 000 years	2.21E+12	8.62E-01	1.00E+04	Matter dominates radiation
1.0 year	3.16E+07	8.62E+01	1.00E+06	Photons dominates growing bubble
1.0 hour	3.60E+03	8.62E+03	1.00E+08	Formation of prime helium
1.0 second	1.00E+00	8.62E+05	1.00E+10	Neutrinos start to escape
1.0 millisecond	1.00E-03	8.62E+07	1.00E+12	Formation of hydrogen nuclei
1.0 microsecond	1.00E-06	4.31E+09	5.00E+13	End of quark domination
1.0 nanosecond	1.00E-09	8.62E+10	1.00E+15	Quark dominates in expanding bubble
1.0 picosecond	1.00E-12	4.31E+12	5.00E+16	Weak and electromag forces separated
1.0 femtosecond	1.00E-15	8.62E+13	1.00E+18	Quark and anti-quark formation
End strong phase	1.00E-35	8.62E+23	1.00E+28	Separation of strong and electroweak
Planck time	1.00E-43	1.22E+28	1.42E+32	Electronuclear rebound on gravity wall

Table 7.1 Bang time sequence for our bubble

The bang time sequence and follow-up events in table 7.1 are transposed from the big bang model to the maxibang at the origin of our northern bubble. The first column indicates the main sequential time scale of events, which is expressed in seconds in column 2. Then related energy and temperature are respectively shown in column 3 and 4, with temperatures varying from 1.4×10^{32} to 2.73 K. Events are shortly described in column 5 from Planck time to present day. In the Planck units defined from five universal physical constants, we also have the Planck mass which represents the maximum mass holding an elementary charge and corresponding to only 2.18X10^{-8} kg, about the mass of the smallest flying insects or of 22,000 amoebas [72].

The first and most remarkable difference from big bang interpretation is that in a maxibang, at Planck time, the electronuclear material emerges from the quasar implosion as a rebound on the gravity wall, and the rebound is not limited by the velocity of EM waves since the emergence is from a different physical domain. So before the electronuclear material escapes, the black force and subnuclear particles dominate the expansion of the shell at a velocity greater than c until a bubble size is reached, where the mass of the shell can escape from collapse at a lower velocity than c. This phenomenon may be compared with the ad hoc flash cosmic inflation in the big bang paradigm; it is a gradual expansion at superluminal speed in the maxibang model. The second remarkable difference is that 13.8 billion years is the age of our own bubble in the multibang model and not the age of the cosmos [83].

As surely as massive stars collapse to neutron stars and black holes, massive clusters of one thousand galaxies plus the equivalent gas mass of six thousand galaxies, likely dressed as quasars with a giant black hole in the center, should collapse all the way to the final force of gravity since all other forces, including the electronuclear and subnuclear black force, have been dissolved and merged under the effect of gravity. Then in a gravity rebound, it is possible to expand faster than the speed of light to a bubble volume and mass where the electronuclear takes over for a standard expansion at a velocity less than c and a shell radius equal to or greater than 5,400 l-y or 1,650 parsecs.

Figure 7.6 Black hole and maxibang mass, volume, and time distribution

The following three graphs are based on one Planck unit only: the Planck length, which is 1.6162×10^{-35} m and is applied as the radius of the smallest and most compact sphere that may be found in nature, so the minimum volume is equal to 1.77×10^{-104} m^3. On all three graphs, we show that black hole compaction proceeds to the right with smaller volumes and that maxibang expansion runs to the left with larger volumes although discontinuities are not inserted in the curves.

The graph of figure 7.6 shows volume and time distribution in the presence of a constant mass of 3.4×10^{44} kg corresponding approximately to the critical mass in our maxibang expansion and contraction model. The horizontal scale for volume is divided in 12 steps of 16 orders of magnitude each, and the logarithmic vertical scale presents SI units for mass, density, and volume.

The square symbols represent bubble volume variations in m^3 over one half cycle. If the first half cycle is bubble expansion to a maximum of 1.77×10^{72} m^3, then the other half is the contraction of material borrowed from common bubble walls and gathered to form a giant cluster of about one thousand galaxies plus massive amount of gas that will be squeezed to the Planck volume. The curve with triangles corresponds to a time of 4.32×10^{17} seconds at the extreme left, which is equal to a bubble age of about 13.7 billion years, the expected age of our bubble. The birth of the electronuclear domain, is shortly followed, as borrowed from the big bang model, by the escape of photons from the growing bubble as shown at 380,000 years. Then time values decrease to zero at the extreme right, where compression reaches the uttermost value at the instant when the bang paroxysm occurs.

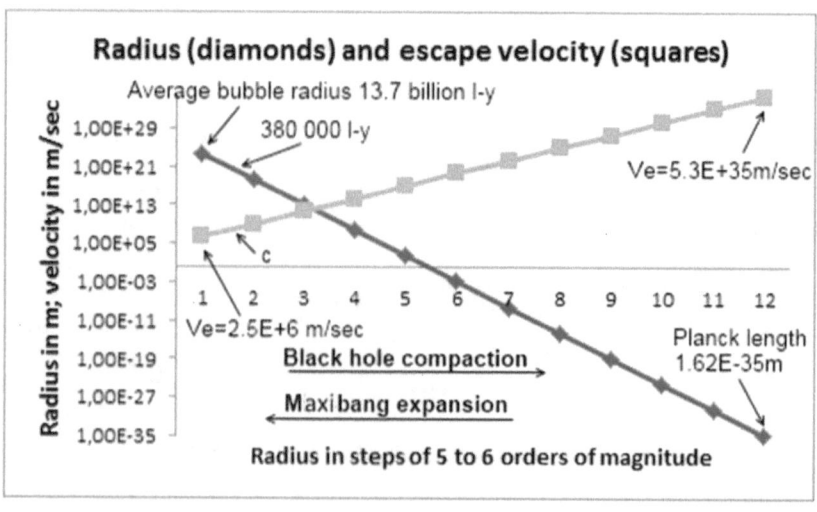

Figure 7.7 Black hole and maxibang radius with escape velocity distribution

In figure 7.7, decreasing radius and increasing escape velocity curves are presented with a horizontal scale divided in 12 steps of five to six orders of magnitude each for radius values shown with diamond symbols. The first value on the left represents an average bubble radius of 7.5×10^{23} m or 13.7 billion l-y, followed by the critical bubble shell radius of 380,000 l-y, marking the end of plasma domination with the formation of neutral hydrogen and the beginning of the free travel of photons away from the expanding bubble. The last value to the right is the Planck radius with a length of 1.62×10^{-35} m. In between the extreme, we find all the intermediate radial values observed and hypothesized during the expansion-contraction cycle.

The escape velocity curve is shown with square symbols and start at the extreme left with a value of 2,500 km/s, which is the average radial expansion velocity for bubbles. It is shortly followed by constant c, the speed of light in a vacuum, and it increases to the extreme right to a maximum value of 5.3×10^{35} m/s at time zero. Of course, such high velocities are in the superluminal range and must be explained. In the contraction phase of the cycle, as soon as massive stars generate black holes, we find part of the material beyond escape velocity c, and as the black holes grow, the subnuclear particles and black forces intervene to form plateaus in black hole contractions [58]. In the expansion phase, the shell of the expanding bubble must reach a radius of 5.07×10^{19} m or about 5,350 l-y before the electronuclear material flashes out of the black force and subnuclear particles.

Figure 7.8 Black hole and maxibang density and temperature distribution

Density and temperature distributions are sketched in figure 7.8, with a horizontal scale divided in 12 steps of five to six orders of magnitude each for the temperature. Black hole compaction runs from left to right and maxibang expansion from right to left. Densities are shown with diamond symbols, which are starting at the extreme left with the average cosmic density of 0.21 M_p/m^3, followed by water and neutron star densities indicated as milestones. The last value to the extreme right is 1.92×10^{148} kg/m³, the density calculated for the critical maxibang mass and the Planck volume.

The curve with square symbols represents temperature variations, with the first value at the left corresponding to the cosmic background temperature at about 2.73 K, which is followed by the average temperature of 3,000 K out of the black age. The last value to the right corresponds to the gravity wall temperature at 2.38×10^{59} K at which all the compacted material from the equivalent of 171 trillion solar masses will rebound to form a new bubble. The black force should be present across most of the curve since densities exceed the escape velocity of the electronuclear material. Note that our original tiny and extremely compact sphere rapidly becomes an expanding shell where high densities and temperatures are maintained in the thin volume of the shell, so we expect some deviations from the projections for both contraction and expansion models.

Figure 7.9 Sketch of a heat pump with cold and hot parts

A sketch of a heat pump is presented in figure 7.9 with its cold and hot sections. The idea is to compare a heat pump cycle to compression and expansion found in a cosmic cycle. In a heat pump, we find a compressor heating up cool gas to a higher hot gas temperature. A heat exchanger at high pressure extracts some of the heat from the gas for heating purposes. The gas is transformed to a colder condensed liquid that is fed into an expansion valve from which it emerges as a cold and lower pressure liquid. The latter is evaporated to form a cool gas that will restart the cycle through the compressor.

Figure 7.10 Sketch of the self-sustained cosmic heat pump

The sketch of the self-sustained cosmic heat pump is presented in figure 7.10, with a transposition of the same elements. The cool gas is found in the small and giant molecular gas clouds generated in bubble expansion and mostly associated with spiral and irregular galaxies. The cool-to-hot gas compressor is the force of gravity that will produce stars (hot gas) from the collapse of protostar eddies and twirls in the cool gas. Hot stars produce a large range of EM waves in their high-pressure interior, including heat released to the surrounding area, in a process similar to that of a heat exchanger. Additional heat is released with cluster formation and black hole compaction while maximum entropy is reached.

The maxibang at time zero with zero entropy is the turning point between the high- and low-pressure sides of the cosmic pump since it marks the start of bubble expansion and cooling accompanied by a gradual increase of the rate of entropy across the whole cycle. The kinetic energy from bubble expansion is transferred to twirls and eddies when the expanding gas meets older walls or other bubbles to form galaxies equivalent to cool gas in a process similar to that of an evaporator.

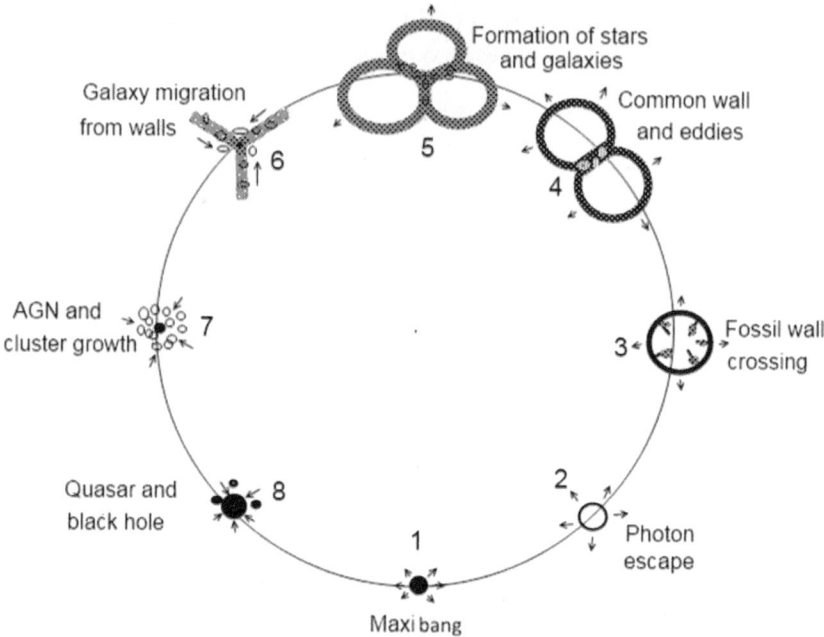

Figure 7.11 One complete cycle from maxibang to black hole

A sketch in eight steps of a complete cycle from maxibang to supergiant black hole with steadily increasing entropy is presented in figure 7.11. The first step is the maxibang event at time zero that is followed by expansion at superluminal speed. A few thousand years later and at sublight speed, photons escape from the expanding shell. In the third step, fossil walls are crossed by the growing bubble that will also, in the fourth step, reach other expanding shells to form common walls and eddies at the origin of stars and galaxies. In the fifth step, the bubble forms up to 12 common walls and up to 200 groups of galaxies with neighbors. Then in step 6, galaxies and gas migrate from the walls to clusters while merging. In step 7, clusters are growing and merging around the increasing mass of a giant central galaxy with a central black hole acting as the dynamo of an active nucleus that will be gradually transformed to a quasar harboring a supergiant black hole in step 8. The supergiant black hole will implode under the force of gravity, and a new cycle will start.

In a complete cycle, black holes are born early from massive stars, and they gradually grow to huge masses within and at the center of galaxies from which they merge into still-larger active-core galaxies and quasars until the final union of about one thousand galaxies plus associated gas that will collapse at a speed of about 5.3×10^{35} m/s in a single black hole that produces the bang.

7.3 Complexity Level and Energy Quality Conservation

In this final portion, we will have a brief look at the development of more and more complex structures that we observe in the cosmos, and we will question the added value of all the new qualities generated at every new level of complexity. In other words, are these qualities part of a new type of force and energy related to some cosmic network based on fundamental laws of nature, or is it some kind of dissipative energy?

7.3.1 Complexity Levels from Gravity to Cosmic Community

So the question is, are we generating conservative or dissipative energy when quarks unite to form protons or neutrons or when molecules unite to form proteins and when organisms unite to form society? It boils down to asking, is the whole greater than the parts when quality and efficiency are increased at a higher complexity level?

Complexity level	Force or bond	Comments
Gravity wall	Gravity	Final encounter at the bang
Subquark	Black	Superluminal and out of experimental reach
Quarks	Strong	Exchange bosons are the gluons
Neutrons/protons	Weak	W and Z bosons and neutrino particles
Nucleus/electrons	Electromagnetic	Electronic charge and spin
Atoms	Chemical bonds	Covalent, ionic, and metallic bonds
Molecules	Molecular bonds	Hydrogen and dipole interactions
Proteins	Peptide bonds	Covalent, hydrogen, and mix bonding
DNA/RNA	Phosphodiester bond	Covalent bond with enzyme-catalyzed reaction
Cells	Biochemical bond	A mixture of weak and strong bonds
Organism	Cellular bond	Cells forming mechanical/ biochemical bonds
Society	Cooperative bond	Group survival by product/info exchange
Cosmic community	Sharing bond	Superluminal exchange and travel??

Table 7.2 Distribution of growing complexity in cosmos

The distribution of growing complexity in the cosmos is presented, from top to bottom, in table 7.2, where the first column is the complexity level, the second column shows the corresponding force or bond, and the third column gives a brief comment. Note that for the second to the fifth complexity levels in the table, a limit of density may be reached where the force is overwhelmed by that of gravity. All the other levels are mainly related to the electromagnetic force subdivided in weaker bonds.

The first level is gravity at the zero and near-zero time when all the future materials of one bubble are accelerated down the final black hole while reaching maximum entropy at a superluminal speed to form a finite "singularity" and rebound on the gravity wall where the entropy is reset at zero. The second level is also in the superluminal range of velocities and out of reach for our limited present-day particle accelerators. It involves the black force acting on subquark particles that could be the hypothetical preon degenerate gas opposing the force of gravity to avoid the collapse of giant quasar black holes. The third level represents the quarks united by the strong force, with the gluons acting as boson exchange particles. But if gravity overwhelms the strong force, we should end up with a degenerate quark gas forming a black hole star with ultrahigh density and a radius equal to or smaller than the Schwarzschild radius from which no light can escape. The latter represents the equivalent of 10 solar masses compacted within a sphere with a radius less than 30 km.

At the fourth level, we find the W and Z bosons as the exchange particles of the weak force for the emission and absorption of neutrinos in and out of neutrons. Again, if gravity takes over the weak force, we may end up with a compacted neutron star having a degenerate neutron gas density in the order of 6×10^{17} kg/m^3. The next level is the electromagnetic force associated with the exchange of photons by electrons and other particles. When the gravitational force overwhelms the electromagnetic force, we end up with a compact star, such as a white dwarf made up of a degenerate electronic to proton gas having a density around one billion kg/m^3.

The following higher complexity levels are all electromagnetic chemical-to-biochemical bonds that represent different levels of resistance against the pull of the relatively weak gravitational force. Between atoms, we find the covalent, ionic, and metallic chemical bonds, which vary in strength. Molecular bonds linking two or more molecules are generally related to hydrogen and dipole interactions. Proteins are formed by peptide bonds where covalent, hydrogen, and mix bonds are present. The DNA and RNA photocopy molecules are formed by phosphodiester bonds, which are covalent bonds catalyzed by an enzyme (protein).

Living cells are born from a mixture of weak and strong biochemical bonds, including proteins and DNA/RNA. Cells are united to form coordinated organs and the moving mechanical parts of an organism. Many individual organisms may join together to form a society where forces or bonds are directed to a cooperative bonding that yields product and information exchange for the survival and well-being of the group. Finally, many planetary or other cosmic societies may unite in order to share information and use common luminal and superluminal exchanges, including transport, for reciprocal betterment.

We know that the entropy of the electronuclear material increases constantly with time even when we go up the rungs of the complexity ladder. Simultaneously with structure increase, we find a general increase in the exchange of information between different units of a specific complexity. It is this data that becomes a plus, contributing to the enrichment of the cosmic network of information. Thus, for a while, some energy is transformed and maintained as bits of qualitative information although at the end, any unit or network is faced with the inescapable increase of entropy.

7.3.2 Quality Increase and Conservation

We will look at quality increase for a specific example when we go from a simple to a more complex level in cosmic bubble evolution. It will be followed by a discussion of what is conserved from all the extra qualities issued from the increased complexity of the cosmos.

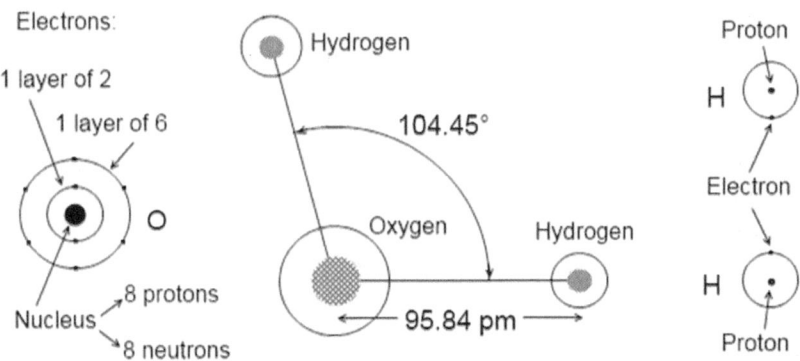

Figure 7.12 Water molecule formed from three atoms

A water molecule formed from three atoms is presented in the center in figure 7.12, with a separate sketch of the two hydrogen atoms on the right and one oxygen atom on the left. The hydrogen atom is the simplest of all atoms and may be represented as a nucleus containing a single proton that is surrounded by an orbiting electron with a charge opposite and equal to that of the proton. When two hydrogen atoms unite, they form molecular hydrogen in which the two electrons are equally shared in a covalent bond in equilibrium with the charge of the two protons.

The oxygen atom has eight protons and eight neutrons held together by the strong force in its nucleus. The charge of the eight protons is neutralized by the opposite charge of two layers of electrons: one inner saturated layer of two electrons and one unsaturated outer layer of six orbiting electrons. Since saturation of the outer layer would require a total of eight electrons, then there is room for sharing this layer with two electrons from other atoms. This is what is represented by the central sketch of a water molecule, where two hydrogen atoms are each sharing their unique electron with an atom of oxygen. A polarized angle of 104.45° is present between the two hydrogen atoms as well as a distance of 95.84 pm between the nucleus of oxygen and any of the two hydrogen nuclei.

The water molecule or hydrogen oxide (H_2O) is found in all regions of the cosmos where temperature permits its formation and survival. It exhibits characteristics that are totally different from its oxygen or hydrogen components. Water has a high melting and boiling temperature as compared with hydrogen or oxygen. It has a hexagonal crystal structure in the frozen phase and a dipole moment producing attraction between individual molecules in the liquid phase. Due to its polarity, water is a good solvent for a large number of substances, especially salts. Using surface tension and adhesion, it can rise against gravity by capillarity in a narrow tube. So water molecules have properties and qualities not found in their basic components, and these new attributes may be questioned for their permanence. Heat, chemical reactions, and electrolysis can destroy a water molecule and bring this compound to its natural entropic death. New H_2O molecules are born by igniting molecular hydrogen H_2 with an O_2 molecule or in a multitude of other reactions. But the lifetime of any individual in the complexity ladder is temporary and bound by its entropic fate just like Humpty Dumpty.

Humpty Dumpty sat on a wall.
Humpty Dumpty had a great fall.
All the king's horses and all the king's men
couldn't put Humpty together again.

7.3.3 Winning Cosmic Mix: Gravity, Black Energy, Electronuclear, and Biochemical

So it seems that our universe is filled to its ultimate limits with a winning cosmic mix made of gravitational, black, electronuclear, and biochemical energy bonds. We have no reason to believe that the laws of nature are different in some region of the cosmos.

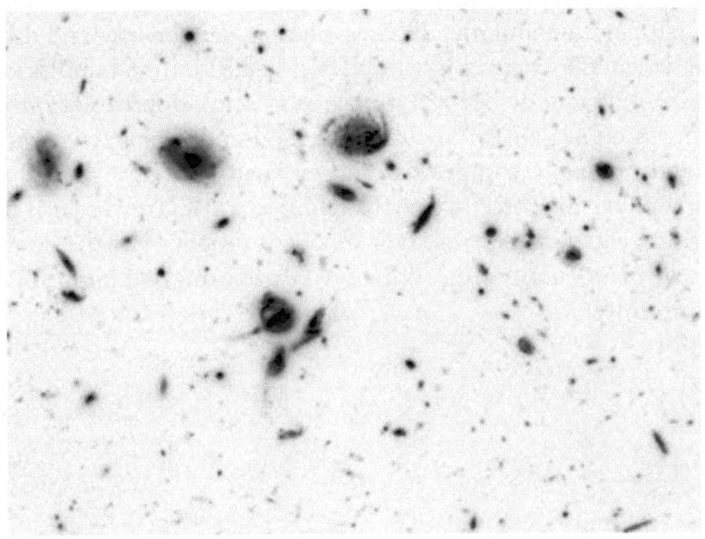

Figure 7.13 A black and white patch of sky from Hubble Extreme Deep Field (Credit: NASA/ESA; G. Illingworth, D. Magee, and P. Oesch, University of California, Santa Cruz; R. Bouwens, Leiden University; and the HUDF09 team)

A small portion of the picture from the Hubble Extreme Deep Field is illustrated in figure 7.13 with black-and-white inverted colors. This little nook of the cosmos is the farthest view ever done from earth in the Fornax constellation. It took 50 days of observation over a period of 10 years and exposure time of two million seconds to assemble the original color image. Thousands of galaxies from nearby and from the very distant universe can be seen and vary from the splendid spirals to the farthest recluses almost the size of a pixel.

In the big bang model, it is estimated that the oldest galaxies may reach an age of 13.2 billion years so that the youngest galaxies observed would be only 450 million years old. How could we find such young galaxies completely formed and mature in such a short period? It is rationally impossible to explain.

In the multibang model, the equivalent maximum age is not limited in time. Thus, when we look at the Hubble Ultra and Extreme Deep Field pictures from NASA, we find, except for the natural filtering from such a large distance, the same old and young galaxies expected from a uniform cosmos made up of a multitude of bubbles with different historical ages. The cosmological principle is saved as an average outlook over very large distances across many bubbles.

Most of the big bang problems [41], like the horizon, flatness, density fluctuation, singularity, structure formation history, galactic and cluster velocities, temperature and metallicity evolution, dark energy, and dark matter, melt away when the cosmos is viewed as a collection of bubbles of various ages that are individually born in a maxibang implosion/explosion whenever a critical mass equivalent to about seven thousand average-size galaxies are gathered and gradually compressed and crushed to a tiny size at zero time and entropy under the force of gravity acting as a clock rewinder. The cycle is unceasingly repeated, while every new version is produced by the homogenized admixture of mass energy from different neighboring bubbles, thus assuring cosmic uniformity.

Summary of Key Ideas in Big Bang (BB) and Multibang (MB) Models

The basic BB premise is anchored in the ingrained idea that the cosmos is a finite anthropocentric object born in time, expanding and eventually dying, not far from mythological stories of creation. In the MB model, the cosmos is a finite and permanent object where the electronuclear material is locally recycled in maxibang explosions and expansions. A flash picture of the cosmos today is just as valid and representative as any flash picture in the past or in the future; the cosmos is permanent.

In both models, redshifts are related to expansion, but the interpretation is quite different. In the BB model, observed redshifts from galaxies surrounding us in our local universe are interpreted as expansion and extrapolated to the whole cosmos as a direct measure of distances via the Hubble constant; spaces with blueshifts or no redshifts are not accounted for and neglected. On the other hand, in the MB model, observed redshifts are associated with those local cells or bubbles that are in expansion, the sum of which adds up to the total redshift and gives an evaluation of distances for very large redshift values; spaces with blueshifts or no redshifts are accounted for and included.

Since the cosmos is expanding equally in all directions in the BB model, then it is possible to go back in time 13.8 billion years ago to the origin of the expansion at time zero where we find an extremely hot and infinitely dense single point called a singularity defying the known laws of physics. So the BB cosmos is gradually hotter and denser in the past. In the permanent MB model, the cosmos always has the same average temperature and density, but every expanding bubble is born in a high-temperature and high-density maxibang explosion that contains a chunk of the cosmos and has its own age and history, representing, for example, about 13.8 billion years for our northern bubble in which the Milky Way is embedded.

In both models, at high temperature and density and shortly after the unique big bang event or after every maxibang events, a reaction called baryogenesis must occur that violates the equilibrium between matter and antimatter, leaving an excess of quarks and leptons that we call matter. Within the first second, protons, neutrons, and electrons are formed and dominated by photons. Then within a few minutes of expansion and at a temperature of about one billion K, a process called nucleosynthesis leaves most of the basic elements in place, including 92% of hydrogen nuclei, 8% of ionized helium, and a small amount of deuterium and lithium.

Finally, less than 400,000 years after the bangs, electrons are gradually combined to helium first and then to hydrogen nuclei to form atoms and release photons that become free to travel across the new gas and the cosmos. In the BB model, the free photons are considered as the unique source of the cosmic microwave background and must travel, mysteriously rebound, and cool down to 2.725 K until they reach us some 13.8 billion years later. But in the MB model, the free photons are constantly emitted by newly born bubbles and mixed with the dominant photonic background from all other sources which, after a long redshifting trip across the cosmos, become the cosmic microwave background.

Many new concepts were invented and developed to fit the BB model as, for example, an ultrarapid expansion, called inflation, at 10^{-35} second of the cosmic birth with the purpose of preserving the cosmological principle of isotropy and homogeneity. In the MB model, every maxibang event is naturally preceded by an implosive giant black hole contraction and followed by an explosive expansion at superluminal speed until it settles to subluminal speed. The cosmological principle is preserved by the equilibrium between bubble expansion and gravity contraction within the cosmos, although local fluctuations are present at scales smaller than z=1.

In the BB model, large structures like bubbles or voids are passively formed by the pull of gravity from the outside, and clusters and groups of galaxies are formed by gravity accretion from and around more massive knots of material, but no satisfactory answer is available on the timing and process. How the huge spherical to polyhedral voids are emptied or how the clusters are built up within the BB time frame of 13.8 G l-y is not accounted for or justified. With a permanent cosmos, the MB model explains the formation of the large voids or bubbles from the inside out by the expansion of maxibang events, the formation of galactic groups and clusters by the dynamic clashing of the shells from the expanding bubbles, and the continuous formation of quasars by the crushing action of giant black holes at the massive galactic center of large clusters. The quasars are eventually imploded-exploded in maxibang events and new bubble expansions.

Based on the apparent velocities of stars and clouds orbiting spiral galaxies and on apparent velocities of galaxies falling in or orbiting clusters of galaxies with associated light-to-mass ratios, it is estimated in the BB model that most of the matter is in the form of dark matter representing 23% of mass-energy, with only about 5% representing ordinary baryonic matter (protons, neutrons). Also, based on a subsequent questionable interpretation of supernovae Ia suggesting accelerated expansion, the BB model includes 72% of the cosmic mass-energy in the form of dark energy with negative pressure. Dark matter and dark energy have never been found, and they don't exist in the MB model where all mass-energy is expected in the form of baryonic matter. In the MB model, corrected disk velocities and not apparent spheroidal velocities are used to measure the orbits of stars around spiral galaxies, thus yielding exact galactic masses and eliminating the fleeing dark matter from the model. A dynamic model of continuous galactic formation and migration to clusters explains the high velocities encountered in cluster suburbs. The naturally larger z distances found in the MB model explain the lower brightness found in high z supernovae Ia surveys. The luminosity-distance curve calculated from supernovae Ia directly corresponds to the MB model distances. So, supernovae Ia are true and reliable standard candles for comic distances. Note that in the MB model, times and distances are presented with the same curve for large distances.

BIBLIOGRAPHY

1. Arp Halton, Catalogue of Discordant Redshift Associations, Montreal, Apeiron, 2003.
2. Arp Halton, Seeing Red, Redshifts, Cosmology and Academic Science, Montreal, Apeiron, 1998.
3. Beer Ferdinand P., Johnston Jr E. Russell, Eisenberg Elliot R., Clausen William E., Vector Mechanics for Engineers, Statics and Dynamics, Seventh Edition, New Delhi, Tata McGraw-Hill Publishing Co Ltd, 2005.
4. Binney James & Tremaine Scott, Galactic Dynamics, Second Edition, Princeton, Princeton University Press, 2008.
5. Carollo C. M., Ferguson H. C. & Wyse R. F. G., Editors, The Formation of Galactic Bulges, Cambridge, Cambridge University Press, 1999.
6. Christensen Lars Lindberg, de Martin Davide, Cosmic Collisions, The Hubble Atlas of Merging Galaxies, New York, Springer Science 2009.
7. Christensen Lars Lindberg, Fosbury Robert, Hurt Robert, Hidden Universe, Weinheim, Wiley-VCH Verlag GmbH & Co., 2009.
8. Cohen-Tannoudji Gilles, Spiro Michel, La Matière-Espace-Temps, La Flèche, Éditions Gallimard, 1990.
9. Diaferio Antonaldo, Editor, Outskirts of Galaxy Clusters: Intense Life in the Suburbs, IAU Colloquium 195, Torino, Italia, Cambridge University Press.
10. Einstein Albert, Conceptions Scientifiques, Traduction de l'anglais, Manchecourt, Flammarion, 2004.
11. Einstein Albert, Infeld Léopold, L'Évolution des Idées en Physique, Traduit de l'anglais, Manchecourt, Flammarion, 2004.
12. Einstein Albert, La Relativité, Traduction de l'allemand, St-Amand, Petite Bibliothèque Payot, 1998.

13. Greene Brian, La Magie du Cosmos, Traduction de l'anglais, Paris, Editions Robert Laffont, 2005.
14. Hoskin Michael, Editor, The Cambridge Concise History of Astronomy, Cambridge, Cambridge University Press, 1999.
15. Kembhavi Ajit K. & Narlika Jayant V., Quasars and Active Galactic Nuclei, An Introduction, Cambridge, Cambridge University Press, 1999.
16. Lerner Eric, The Big Bang Never Happened, New York, Vintage Books, 1992.
17. Luminet Jean-Pierre, L'Univers chiffonné, St-Amand, Gallimard, 2007.
18. Mészaros Péter, The High Energy Universe, Ultra-High Energy Events in Astrophysics and Cosmology, Cambridge, Cambridge University Press, 2010.
19. Pascoli Gianni, La Gravitation, Paris, Presses Universitaires de France, 2nd edition, 1995.
20. Ratcliffe Hilton, The Static Universe, Exploding the Myth of Cosmic Expansion, Montreal, Apeiron, 2010.
21. Reeves Hubert, Dernières Nouvelles du Cosmos, Paris, Éditions du Seuil, 1994.
22. Reid I. Neill & Hawley Suzanne L., New Light on Dark Stars, Red Dwarfs, Low-Mass Stars, Brown Dwarfs, Chichester, Springer-Praxis, 2005.
23. Schneider Peter, Extragalactic Astronomy and Cosmology, An Introduction, Berlin Heidelberg, Springer-Verlag, 2006.
24. Séguin Marc, Villeneuve Benoît, Astronomie & Astrophysique, St-Laurent, Editions du Renouveau Pédagogique Inc., 1995.
25. Serway Raymond A., Physics For Scientists & Engineers with Modern Physics, Third Edition, Philadelphia, Saunders College Publishing, 1990.

Internet Articles

26. Arp Halton, What Is the Mean Redshift of the Virgo Cluster?, 1988, Astronomy & Astrophysics, **202**, 70–76.
27. Ashmore Lyndon, An Explanation of Redshift in a Static Universe, 2010, NPA, **6**(2), 1–6.
28. Briggs Michael, Molecular Hydrogen Excitation in Gas Clouds, 2007, *www.roe.ac.uk/ifa/postgrad/pedagogy/2007_briggs.pdf*.
29. Brownstein J. R., Moffat J. W., Galaxy Rotation Curves without Non-Baryonic Dark Matter, 2005, arXiv:astro-ph/0506370v4.
30. Bruchon J., Fortin A., Bousmina M., Benmoussa K., Direct 2D Simulation of Small Bubble Clusters: From the Expansion Step to the Equilibrium State, 2006, International Journal for Numerical Methods in Fluids, **00**, 1–29.
31. Courteau Stéphane, Faber S. M., Dressler Alan, Willick Jeffrey A., Streaming Motions in the Local Universe: Evidence for Large-Scale, Low-Amplitude Density Fluctuations, 1993, The Astrophysical Journal, **412**, L51–L54.
32. De Blok W. J. G., Bosma A., High-resolution Rotation Curves of Low Surface Brightness Galaxies, 2002, Astronomy & Astrophysics, **385**, 816–846, *http://www.aanda.org*.
33. Folkman Judah, Hochberg Mark, Self-Regulation of Growth in Three Dimensions, 1973, The Journal of Experimental Medicine, **138**, 745–753.
34. Fraser Cain, Clouds of Hydrogen Swarm around Andromeda, 2004, NRAO News Release.
35. Galactic Rotation Curves, *http://cosmology.berkeley.edu/Education/Essays/galrotcurve.html*.
36. Gregory Stephen A., Thomson Laird A., The Coma/A1367 Supercluster and Its Environs, 1978, Astrophysical Journal, **222**, 784–799.

37. Henry Richard C., Diffuse Background Radiation, 1999, The Astrophysical Journal, 516, L49–L52.
38. Kassin Susan A., de Jong Roelof S., Weiner Benjamin J., Dark and Baryonic Matter in Bright Spiral Galaxies: II. Radial Distributions for 34 Galaxies, 2006, arXiv:astro-ph/0602027 v1.
39. Marmet L., Gallo C. F., Comment on the Use of Kepler's Laws to Describe Galactic Rotation, *http://www.marmet.org/cosmology/nonkeplerian/galaxy.html*.
40. McGaugh Stacy S., Boomerang Data Suggest a Purely Baryonic Universe, 2000, The Astrophysical Journal, **541**, L33–L36.
41. Meta Research, The Top 30 Problems with the Big Bang, 2002, Meta Research Bulletin, **11**, 6–13, *http://metaresearch.org/cosmology/BB-top-30.asp*.
42. Ofer Yaron, The Galactic Environment of the Sun, *http://www.tau.ac.il/~oferya/documents/The%20Galactic%20Environment%20of%20The%20Sun%20final.pdf*.
43. Pan Danny C., et al., Cosmic Void in Sloan Digital Sky Survey Data Release 7, 2011, arXiv:1103.4156v2 [astro-ph.CO].
44. Percival Will J., et al., The 2dF Galaxy Redshift Survey: The Power Spectrum and the Matter Content of the Universe, 2001, Mon. Not. R. Astron. Soc., **327**, 1297–1306.
45. Savin D. W. et al., The Impact of Recent Advances in Laboratory Astrophysics on Our Understanding of the Cosmos, 2012, IOP Publishing, Reports on Progress in Physics, **75**, 1–36.
46. Shiga David, Dwarf-flinging Void Is Larger than Thought, New Scientist, 2007.
47. The Andromeda Galaxy Seen as Never, at the 21-cm Wavelength, 2009, *http://www.obspm.fr/actual/nouvelle/oct09/m31.en.shtml*.
48. Types of Galaxies, *http://www.universe-review.ca/F05-galaxy.htm*.
49. Van de Weygaert, et al., The Void Galaxy Survey, 2011, arXiv:1101.4187v1 [astro-ph.CO].
50. Van Flanders Tom, Vigier J. P., The Speed of Gravity-Repeal of the Speed Limit, Meta Research, 2002, *http://metaresearch.org/cosmology/gravity/speed_limit.asp*.
51. Weiss Achim, Elements of the Past: Big Bang Nucleosynthesis and Observation, 2006, Einstein Online, **2**, 1019.
52. Wikipedia, The Free Encyclopedia, Big bang, *http://en.wikipedia.org/wiki/Big_Bang*.
53. Wikipedia, The Free Encyclopedia, Black hole, *http://en.wikipedia.org/wiki/Black_hole*.

54. Wikipedia, The Free Encyclopedia, Cosmic distance ladder, http://en.wikipedia.org/wiki/Extragalactic_distance_scale.
55. Wikipedia, The Free Encyclopedia, Cosmic microwave background radiation, http://en.wikipedia.org/wiki/Cosmic_microwave_background_radiation.
56. Wikipedia, The Free Encyclopedia, Dark energy, http://en.wikipedia.org/wiki/Dark_energy.
57. Wikipedia, The Free Encyclopedia, Dark matter, http://en.wikipedia.org/wiki/Dark_matter.
58. Wikipedia, The Free Encyclopedia, Degenerate matter, http://en.wikipedia.org/wiki/Degenerate_matter.
59. Wikipedia, The Free Encyclopedia, Doppler effect, http://en.wikipedia.org/wiki/Doppler_effect.
60. Wikipedia, The Free Encyclopedia, Electromagnetic spectrum, http://en.wikipedia.org/wiki/Electromagnetic_spectrum.
61. Wikipedia, The Free Encyclopedia, Extragalactic background light, http://en.wikipedia.org/wiki/Extragalactic_background_light.
62. Wikipedia, The Free Encyclopedia, Friedmann equations, http://en.wikipedia.org/wiki/Critical_density.
63. Wikipedia, The Free Encyclopedia, Interstellar medium, http://en.wikipedia.org/wiki/Interstellar_medium.
64. Wikipedia, The Free Encyclopedia, Kepler conjecture, http://en.wikipedia.org/wiki/Kepler_conjecture.
65. Wikipedia, The Free Encyclopedia, List of voids, http://en.wikipedia.org/wiki/List_of_voids.
66. Wikipedia, The Free Encyclopedia, Local void, http://en.wikipedia.org/wiki/Local_Void.
67. Wikipedia, The Free Encyclopedia, Lyman-alpha forest, http://en.wikipedia.org/wiki/Lyman-alpha_forest.
68. Wikipedia, The Free Encyclopedia, Main sequence, http://en.wikipedia.org/wiki/Main_sequence.
69. Wikipedia, The Free Encyclopedia, Molecular cloud, http://en.wikipedia.org/wiki/Molecular_cloud.
70. Wikipedia, The Free Encyclopedia, Neutron star, http://en.wikipedia.org/wiki/Neutron_star.
71. Wikipedia, The Free Encyclopedia, Physics beyond the Standard Model, http://en.wikipedia.org/wiki/Physics_beyond_the_standard_model.
72. Wikipedia, The Free Encyclopedia, Planck units, http://en.wikipedia.org/wiki/Planck_units.
73. Wikipedia, The Free Encyclopedia, Quasar, http://en.wikipedia.org/wiki/Quasar.

74. Wikipedia, The Free Encyclopedia, Redshift, *http://en.wikipedia.org/wiki/Redshift*.
75. Wikipedia, The Free Encyclopedia, Soap bubble, *http://en.wikipedia.org/wiki/Soap_bubble*.
76. Wikipedia, The Free Encyclopedia, Spiral galaxy, *http://en.wikipedia.org/wiki/Spiral_galaxies*.
77. Wikipedia, The Free Encyclopedia, Stellar classification, *http://en.wikipedia.org/wiki/Stellar_classification*.
78. Wikipedia, The Free Encyclopedia, Supernova, *http://en.wikipedia.org/wiki/Supernova*.
79. Wikipedia, The Free Encyclopedia, Theory of everything, *http://en.wikipedia.org/wiki/Theory_of_everything*.
80. Wikipedia, The Free Encyclopedia, Tully-Fisher relation, *http://en.wikipedia.org/wiki/Tully-Fisher_relation*.
81. Wikipedia, The Free Encyclopedia, Vacuum, *http://en.wikipedia.org/wiki/Vacuum*.
82. WMAP, What Is the Universe Made Of? map.gsfc.nasa.gov/.../uni_matter.html.
83. Wright Edward L., Age of the Universe, 2009, Tutorial/Age.
84. Gallo C. F., Feng James Q., Galactic Rotation Described by a Thin-Disk Gravitational Model without Dark Matter, 2010, Journal of Cosmology, **6**, 1373–1380.
85. Springel Volker, Hernquist Lars, The History of Star Formation in a Lambda Cold Dark Matter Universe, 2003, Monthly Notice of the Royal Astronomical Society, **339**(2), 312–334.
86. Bechtold J., Dobrzycki A., Wilden B., Morita M., Scott J., Dobrzycka D., Tran K., Aldcroft T. L., A Uniform Analysis of the Lyman Alpha Forest at z=0–5: III. HST FOS Spectral Atlas, 2001, ApJS, in press. NSF and NASA.
87. Bechtold J., The Lyman Alpha Forest Near 34 Quasars with z>2.6, 1994, ApJ Supplement, 91, NSF and NASA.
88. Crawford David F., Observations of Type 1a Supernovae Are Consistent with a Static Universe. arXiv:0901.4172v1 [astro-ph.CO] 27 Jan 2009.
89. Hawkins M. R. S., On Time Dilation in Quasar Light Curves. arXiv:1004.1824v1 [astro-ph.CO] 11 Apr 2010.
90. Hurier G. et al., Measurement of the T_{CMB} Evolution from the Sunyaev-Zel'dovich Effect, 2013, Astronomy & Astrophysics manuscript no TCMB. arXiv:1311.4694v3 [astro-ph.CO] 10 Dec 2013.
91. Wikipedia, The Free Encyclopedia, Metallicity, *http://en.wikipedia.org/wiki/Metallicity*

92. Mannucci et al., A Fundamental Relation Between Mass, Star Formation Rate and Metallicity in Local and High-redshift Galaxies, 2010, MNRSA **408** (4), 2115–2127.
93. Pettini Max et al., The Metallicity of High-Redshift Galaxies: The Abundance of Zinc in 34 Damped Ly alpha Systems from z=0.7 to 3.4, 1997, The Astrophysical Journal, 486, 665–680.

Index

A

absorption troughs, 107, 143–44
acceleration, 19, 44, 98, 100–101, 138, 167
accretion, 67, 120–21, 129, 168
accretion disk, 118
active galactic nucleus (AGN) galaxies, 25, 67, 88, 101, 105, 130–31
air, 33, 36, 38, 49–50, 52, 56–57
amplitude, 49, 143, 145
  relative, 144, 146
An-1, 155–56
anomalies, secondary, 102
antiparticle, 211
apparent velocities, 25, 233
Ar-1, 155
astrophysics, 48, 51, 112, 209, 213
atmosphere, 11, 115
atoms, 14, 22, 24, 36, 112, 115, 143, 226–28

attraction, 14, 150, 160–61, 163–65, 167–68, 173, 177–78, 190, 198–203
attraction radius, 165, 173
average density, 14, 24, 36–37, 132–33, 136, 196
average distance, 109
average light trajectory, 74
average mass, 66, 132–33, 136–37, 188
average radius, 26, 37, 45, 63, 66, 132, 136
average redshifts, 23, 79
average spectrum, 12, 146
average spherical radius, 133
average spiraling, 14, 197
axis, 127

B

bang, 23, 31, 93, 102, 123–24, 141, 224–25, 232
bang time sequence, 9, 59, 218

barred galaxies, 128, 184, 187–88, 195–96
barred spiral, 131, 134
barycenter, 150, 199
  common, 117, 130
barycenters, secondary, 130
baryon, 39, 44
baryon density, 10, 38–39, 43
baryonic matter, 19, 24, 101–2, 233
BB cosmos, 231
BB model, 231–33
big bang cosmic expansion model, 74
big bang cosmological models, 44
big bang creation, 24
big bang curves, 101
big bang estimates, 40
big bang interpretation, 219
big bang model, 19, 21, 23, 38, 40–41, 44–45, 54, 65, 74, 94, 97–101, 122–23, 140, 146–47, 165, 179, 212, 218, 220, 229
  compact, 100
  prevalent, 17
  unique, 18
big bang model calculations, 183
big bang paradigm, 20–21, 23, 99, 219
big bang proponents, 22, 45, 107, 217
big bang reionization model, 140
big bang theory, 21
blackbody spectra, 11, 102–3
black energy, 132, 229
black hole center, active, 138
black holes, 14, 18, 37, 41, 67, 83, 101, 111, 116, 118, 121, 127, 129–32, 181, 184, 187–88, 212, 214, 217, 219–21, 224, 226, 233
  massive, 131, 209
blueshifting bubbles, 61, 86
blueshifts, 23, 49, 51, 59, 85, 96, 231
bosons, 14, 209–11, 225–26
brightness, 105, 108, 147
bubble cycles, 117
bubble density, 43, 66
bubble diameter, 63, 74
bubble expansion, 10, 25, 38, 59, 69–70, 82, 85, 89–90, 98, 122, 211, 216, 220, 223, 232
  multiple, 59
  redirected, 84
bubble expansion velocity, 64
bubble expansion velocity range, 70
bubble pressure, 41
bubble radius, 33, 45, 70, 109
bubbles, 9–11, 18, 23, 25–26, 32–38, 41–43, 45, 52, 54, 58, 60–74, 76–80, 82–83, 85–87, 89–93, 96–98, 101–2, 109, 122–26, 132–34, 136–38, 141, 143, 218–21, 230–32
  active, 124, 137
  birth-giving, 77
  born, 232
  central, 71, 86
  compacted, 87
  equal-surface, 34
  gaseous, 96
  growing, 68, 218, 220, 224
  local cosmic, 77
  low-density, 36, 57, 71
  neighboring, 61, 68, 74, 82, 230
  neutral nonexpanding, 64
  new, 41, 69, 71, 93, 222

old, 92–93
southern, 37
spheroidal, 63
bubble shells, 84
bulge, 19, 127, 129, 135, 171, 176, 180, 184, 186–87, 198, 200
  elongated, 167–68, 178, 186, 202
  massive, 195–96

C

candles, standard, 81, 94, 99–100, 121
cells, 18, 26, 41, 95–96, 117, 225–26
cellular expansion model, 95
Centaurus cluster, 78
Centaurus supercluster, 78, 80–81, 84
center of attraction, 150, 160, 163–65, 168, 173, 177–78, 199–201
center of mass, 12, 149–50, 163–65, 173, 177–78, 198–201, 204
central spheroid model, 180
Cepheid variable stars, 20
Chandrasekhar limit, 118
circular velocity equation, 165, 169–70, 172–73, 177, 179, 205
clouds, 15, 26, 45, 58, 111, 117, 121–23, 126, 128, 135, 138, 140, 203
clusters, 10–12, 25–26, 31, 36–37, 39, 41, 45, 52, 61–64, 66–67, 77–78, 84–85, 88–90, 97–98, 102, 106, 109, 111, 122–23, 129, 136–38, 203, 207–8, 224, 232–33

Cosine-Corrected Gravitational Force, 12, 151–53, 158, 161
Cosmic Background Explorer (COBE), 104
cosmic bubble number, 9, 66
cosmic infrared background (CIB), 105
cosmic light redshift, 53
cosmic mass, 23, 42–43, 71
cosmic mass distribution, 25
cosmic mass energy, 101
cosmic microwave background (CMB), 18, 22–23, 25, 39, 43–44, 56, 59, 84, 101–9, 113–14, 232
cosmic radio background (CRB), 105
cosmic radius, 40, 45, 54, 109
cosmic redshift, 23
cosmological Doppler redshift, 24
cosmological principle, 22, 230, 232
cosmological principle of homogeneity, 23–24
cosmologists, 61
cosmology, 17
cosmos, 9, 11, 17–19, 21–26, 30, 32, 38–45, 52, 58–60, 64, 66, 72, 74, 86, 88, 90, 93, 96, 101, 105, 107–9, 139–42, 209–13, 215–17, 225–32
  dead, 41
  evaporating, 40
  expanding, 102
  finite, 36
  finite spheroidal, 18
  isotropic, 45
  local, 122
  permanent, 41, 232
  proximal, 82

spheroidal multibang, 42
static, 44
uniform, 23, 230
young, 23
covalent bonds, 225–26, 228
crossover, 167–68, 176, 178, 182–83, 185–87, 189–90, 202, 205–6
cycles, complete, 14, 48, 51, 224

D

damped Lyman alpha (DLA), 122
dark energy, 19, 25, 44, 97, 99–101, 230, 233
   exotic, 23
   hypothetical, 41, 100
dark matter, 19, 89, 97, 99, 101, 178, 180, 183, 190, 230, 233
   nonbaryonic, 23
diamonds, 12–13, 161–63, 169, 182, 185, 189, 201, 214
diamond symbols, 176, 197–98, 221–22
   black, 194, 196
disk equations, 155
dodecahedron, 11, 30, 92
Doppler effect, 10, 46, 48–50, 143
dwarf
   red, 113, 133
   white, 15, 118–19, 121, 226
dwarf stars, 15, 118

E

effects, gravitational disk, 179

electronuclear energy, 41, 101, 212, 214, 216
elliptical galaxies, 13, 94, 123, 129–30, 135, 137, 173, 188, 191
   dwarf, 129
   elongated, 131
   spheroidal, 188
energy, 18, 21–22, 24–25, 27, 36–37, 40–41, 44, 46, 48, 51, 55–56, 101–3, 105, 107, 110, 117–19, 121, 128, 131, 143, 209–18, 225, 227
entropy, 209, 213–14, 223, 226–27, 230
escape velocity, 39, 221–22
exchange bosons, 210–11, 225
exchange particles, 211, 226
expansion, 10–11, 18, 20–21, 24, 38, 41, 44–45, 52–53, 59, 61, 63–66, 68, 73–74, 77, 79, 81–82, 84–86, 93, 95, 98, 100, 123, 133, 138, 231–32
   accelerated, 19, 23, 25, 233
   cosmic distance, 97
   new bubble, 233
expansion energy, 123
expansion models, 222

F

face-centered cubic arrangement, 29
fermions, 14, 209–11
field galaxies, 71
   solitary, 132
filamentary structures, 26, 30, 37, 57–58, 66, 71, 85, 123, 203, 207
filaments, 25, 30–31, 135
flux, 143, 146–48

relative, 144–46
frequencies, 46
frequency scales, 103–4
friction, 49, 209, 213–14
fusion, 22, 113, 118, 121, 128

G

galactic center, 135, 202
  massive, 233
galactic clusters, 69, 77, 82, 105
galactic disk, 149, 151, 164, 166–67, 173, 177, 180–81, 184, 187, 191, 196, 199
  homogeneous, 166
galactic disk computations, 172
galactic disk radius, 197
galactic evolution, 12, 131, 138
galactic formation, 83–84
galactic groups, 80, 88, 133, 203, 232
galactic masses, 128, 157, 165
galaxies, 10–11, 14, 19–20, 25–26, 29–32, 36–37, 60–62, 64–71, 82–84, 86, 89–91, 111, 116–19, 122–30, 132–38, 149, 151–52, 168–69, 175–78, 181–85, 187–88, 190–203, 224, 229, 231–33
  merged, 116, 129
  satellite, 15, 130
  void, 10, 70–72, 122
gamma rays, 48, 67, 104, 118, 121, 138
gaseous spheres, 31
gluons, 210–11, 225–26
gravitation, 14, 176, 209, 216

gravitational field, 18, 44, 128, 130, 132, 150
gravity, 18–19, 24, 41, 44–45, 67, 95, 101, 111, 117–18, 121, 123, 126, 128, 132, 150, 214–17, 219, 223–26, 228–30, 232
gravity wall, 37, 212, 218–19, 225–26
growing complexity, 9, 225–26

H

Hertzsprung-Russell diagrams, 94
hexahedrons, 10–11, 31, 70, 89–92, 141
  concave, 11, 83, 90–91
Higgs boson, 211
high-surface-brightness (HSB) galaxies, 122, 175, 192, 197
Hubble rate, 70, 81–82

I

infinite density, 22, 131
inverse compton scattering, 101, 106, 108–9
inverse laws, 12, 147
ionized atoms, 123
ionized gas, 126, 135
irregular galaxies, 126, 130, 223
  numerous, 130
isotropy, 18, 23–24, 41, 232

K

Kepler conjecture, 29, 89, 92–93
Keplerian mass, 170, 207
kinetic energy, 86, 90, 123–24, 126–30, 135, 168, 214–15, 223
  high, 84, 93

L

laws of thermodynamics, 14, 213
lookback time, 11, 97
low redshift, 61, 86, 122, 203
low-surface-brightness (LSB) galaxies, 192–98
Lyman alpha, 143–44, 146
Lyman forest, 144–46

M

mare incognitum, 212
mass, 11–14, 18–19, 37, 39, 44–45, 66, 112–14, 117–18, 120–21, 123, 128–29, 132, 135–37, 141–42, 149–52, 156–65, 167–79, 181, 183–84, 186–88, 190–91, 194–95, 198–204, 210–11, 219–20
  apparent, 13, 169, 171, 178–79, 183, 187, 191, 206–7
  calibrated, 169–71
  central, 40, 89, 111, 117, 138, 191
  compacted, 37
  dense, 132
mass value, 171
mass values, apparent, 169, 171, 179

maxibang density, 14, 221
maxibang events, 19, 37, 45, 92, 142, 211, 213, 217, 224, 232–33
maxibang expansion, 220, 222
maxibang mass, 14, 219
  critical, 222
maxibang model, 212, 219
maxibang radius, 14, 220
maxibangs, 9–10, 14, 25, 38, 45, 66, 68, 83, 93, 138, 141–42, 218–19, 223–24
  high-density, 231
merging galaxies, 203, 208
metallicity, 23, 122
  high, 122
  low, 122
Milky Way, 15, 20, 37, 76, 80, 84, 113, 116, 118, 123, 130, 132–34, 143, 180, 202, 232
molecular clouds, 77, 117, 121, 128, 130, 135
molecular hydrogen, 116, 123–25, 135, 168, 228
molecules, 24, 115, 123–24, 225–26, 228

N

nebulae, 20
neutrinos, 22, 37, 39, 57, 111, 121, 210, 216, 226
neutrons, 22, 57, 121, 216, 225–26, 228, 232–33
neutron stars, 15, 37, 118–19, 121, 219
  compacted, 226

massive, 216
Newton, Isaac, 12–13, 25, 39, 149, 151, 158, 161–62, 164, 173, 175–76, 180, 182, 185, 189, 200, 216
   universal law of gravitation, 176
   universal law of gravitational attraction, 25
novae, 118
nucleosynthesis, 10, 21–22, 25, 38, 43, 232

P

photons, 11, 18, 24, 39–42, 44, 58, 97–98, 101, 106, 108–9, 111, 140, 142–43, 145–46, 210–11, 216, 218, 220–21, 226, 232
   energetic, 134
   high-energy, 135
   low-energy, 105
   redshifted, 108, 142
Planck epoch, 22
Planck mass, 219
pressure, 33, 69–70, 88, 168
prism, 10, 48, 57
projections, 10, 14, 62, 140, 146, 212, 222
proton masses, 37, 43, 66, 71, 132–33, 136
protons, 22, 38, 138, 217, 228, 232–33

Q

quasar distribution, 139

quasars, 9–10, 12, 19, 54, 67–68, 72–73, 80, 83, 90, 92–93, 100–101, 109, 123, 130–31, 136–43, 145–47, 213, 219, 224, 232–33
   redshifted, 145
quasar spectra, 98, 147

R

radiations, 22, 115, 213, 218
   high-energy, 117, 123
radio wavelengths, 103, 108–9, 130
radius, 9–10, 12, 28–30, 33–34, 37, 39–41, 63, 73–74, 109, 112–14, 127–28, 151–52, 157, 161–63, 169, 171, 173, 175–76, 179, 181, 183, 193–95, 198–202, 220–21, 226
redshift, 9–11, 20–21, 23, 50–55, 59–62, 64–65, 73–74, 81, 84, 94–100, 106–10, 122, 139–40, 142–48, 231
refraction, 10, 25, 42, 46, 48, 56–58
relative energy density, 114
relativity, 22–23, 44

S

Schwarzschild equation, 39–40
Schwarzschild radius, 18, 226
Sloan Digital Sky Survey (SDSS), 59, 94
spheroid, 11, 13, 62, 96, 129, 180, 184, 191, 216
spheroidal bulges, 127–28
   larger, 128

spheroidal model, 190, 195
  permanent, 41
spheroidal velocities, apparent, 233
star formation, 116, 126–28, 135, 202
  massive, 126
star globular clusters, 124
stars, 9, 11, 19, 22, 24, 26, 36–39, 41, 45, 68, 94, 105, 107, 111–24, 128, 130, 132–33, 135–38, 140, 175, 202–3, 213, 217, 223–24, 233
  black hole, 226
  companion, 121
  dead, 120
  discrete, 123
  explosive, 94, 128
Sunyaev-Zel'dovich effect, 11, 59, 101–2, 105–6, 109
superclusters, 203
supernovae, 19, 44, 81, 94, 98–101, 118, 120–21, 233
  distant, 101
supernova redshift, 23

T

tetrahedrons, 30, 35, 87, 91–92
  contracting, 11, 88
theory of relativity, 23, 44

V

vectors, 12, 19, 158–59, 161–62, 167, 186, 197
velocities, 10, 13–14, 19, 23, 39, 48, 61, 69–70, 79–81, 83–84, 123, 128–29, 135, 156–57, 163–69, 172, 174, 177–79, 186, 190, 197, 201–2, 204–5, 215, 219
  angular, 127, 202
  cosine-corrected, 157, 170
  high, 89, 130, 135, 138, 221, 233
  lower, 83, 219
Vizier catalogue, 12, 139
volume-density equation, 157

W

wavelengths, 11–12, 46, 48–50, 52, 54–55, 57–58, 95, 103–4, 106, 109–10, 114, 133–34, 143–46
  redshifted, 142–43
Wilkinson Microwave Anisotropy Probe (WMAP), 104

Z

zircon, 56–57

www.ingramcontent.com/pod-product-compliance
Lightning Source LLC
Chambersburg PA
CBHW020742180526
45163CB00001B/314